Diesel Engine Handbook

Diesel Engine Handbook

Edited by **Nicole Maden**

CLANRYE
INTERNATIONAL

New Jersey

Published by Clanrye International,
55 Van Reypen Street,
Jersey City, NJ 07306, USA
www.clanryeinternational.com

Diesel Engine Handbook
Edited by Nicole Maden

International Standard Book Number: 978-1-63240-144-1 (Hardback)

Contents

Preface

Diesel engine is acknowledged for its superior efficiency and possesses a wide field of applications. It is also known as CI engine. Diesel engines also however, are the prime source of emissions such as NOX and particulate matter (PM). In order to reduce the emissions to an absolute minimum, this book explain as to how these toxins can be regulated. It is no hidden secret that the world is witnessing an oil crisis. But with other alternative sources such as biogas, natural gas and coke based substances; diesel is not the only way forward. The unique characteristics and properties such as combustion and emission of the aforementioned alternatives are explained extensively in this book. The book also goes on to explain how one can look for early signs of wear and tear and malfunctioning components of a diesel engine and its parts.

The information shared in this book is based on empirical researches made by veterans in this field of study. The elaborative information provided in this book will help the readers further their scope of knowledge leading to advancements in this field.

Finally, I would like to thank my fellow researchers who gave constructive feedback and my family members who supported me at every step of my research.

Editor

Combustion and Emissions

The Effect of Split Injection on the Combustion and Emissions in DI and IDI Diesel Engines

S. Jafarmadar

Additional information is available at the end of the chapter

1. Introduction

The major pollutants from diesel engines are NOx and soot. NOx and soot emissions are of concerns to the international community. They have been judged to pose a lung cancer hazard for humans as well as elevating the risk of non-cancer respiratory ailments. These emissions react in the atmosphere in the presence of sunlight to form ground-level ozone. Ground-level ozone is a major component of smog in our cities and in many rural areas as well. In addition, NOx reacts with water, oxygen and oxidants in the atmosphere to form acid rain. Furthermore, the indirect effect of NOx emission to global warming should be noted. It is possible that NOx emission causes an increase secondary emission formation and global warming.

Stringent exhaust emission standards require the simultaneous reduction of soot and NOx for diesel engines, however it seems to be very difficult to reduce NOx emission without increasing soot emission by injection timing. The reason is that there always is a contradiction between NOx and soot emissions when the injection timing is retarded or advanced.

Split injection has been shown to be a powerful tool to simultaneously reduce soot and NOx emissions for DI and IDI diesel engines when the injection timing is optimized. It is defined as splitting the main single injection profile in two or more injection pulses with definite delay dwell between the injections. However, an optimum injection scheme of split injection for DI and IDI diesel engines has been always under investigation.

Generally, the exhaust of IDI diesel engines because of high turbulence intensity is less smoky when compared to DI diesel engines [1]. Hence, investigation the effect of split injection on combustion process and pollution reduction of IDI diesel engines can be quite valuable.

In an IDI diesel engine, the combustion chamber is divided into the pre-chamber and the main chamber, which are linked by a throat. The pre-chamber approximately contains 50% of the combustion volume when the piston is at TDC. This geometrical represents an additional difficulty to those deals with in the DI combustion chambers. Fuel injects into the pre-combustion chamber and air is pushed through the narrow passage during the compression stroke and becomes turbulent within the pre-chamber. This narrow passage speeds up the expanding gases more.

In the recent years, the main studies about the effect of the split injection on the combustion process and pollution of DI and IDI diesel engines are as follows.

Bianchi et al [2] investigated the capability of split injection in reducing NOx and soot emissions of HSDI Diesel engines by CFD code KIVA-III. Computational results indicate that split injection is very effective in reducing NOx, while soot reduction is related to a better use of the oxygen available in the combustion chamber.

Seshasai Srinivasan et al [3] studied the impact of combustion variations such as EGR (exhaust gas recirculation) and split injection in a turbo-charged DI diesel engine by an Adaptive Gradient-Based Algorithm. The predicted values by the modeling, showed a good agreement with the experimental data. The best case showed that the nitric oxide and the particulates could be reduced by over 83% and almost 24%, respectively while maintaining a reasonable value of specfic fuel consumption.

Shayler and Ng [4] used the KIVA-III to investigate the influence of mass ratio of two plus injections and delay dwell on NOx and soot emissions. Numerical conclusions showed that when delay dwell is small, soot is lowered but NOx is increased. In addition, when delay dwell is large, the second injection has very little influence on soot production and oxidation associated with the first injection.

Chryssakis et al [5] studied the effect of multiple injections on combustion process and emissions of a DI diesel engine by using the multidimensional code KIVAIII. The results indicated that employing a post-injection combined with a pilot injection results in reduced soot formation; while the NOx concentration is maintained at low levels.

Lechner et al [6] analyzed the effect of spray cone angle and advanced injection-timing strategy to achieve partially premixed compression ignition (PCI) in a DI diesel engine. The authors proved that low flow rate of the fuel; 60-degree spray cone angle injector strategy, optimized EGR and split injection strategy could reduce the engine NOx emission by 82% and particular matter by 39%.

Ehleskog [7] investigated the effect of split injection on the emission formation and engine performance of a heavy-duty DI diesel engine by KIVA-III code. The results revealed that reductions in NOx emissions and brake-specific fuel consumption were achieved for short dwell times whereas they both were increased when the dwell time was prolonged.

Sun and Reitz [8] studied the combustion and emission of a heavy-duty DI diesel engine by multi-dimensional Computational Fluid Dynamics (CFD) code with detailed chemistry, the

KIVA-CHEMKIN. The results showed that the start of late injection timing in two-stage combustion in combination with late IVC timing and medium EGR level was able to achieve low engine-out emissions.

Verbiezen et al [9] investigated the effect of injection timing and split injection on NOx concentration in a DI diesel engine experimentally. The results showed that advancing the injection timing causes NOx increase. Also, maximum rate of heat release is significantly reduced by the split injection. Hence, NOx is reduced significantly.

Abdullah et al [10] progressed an experimental research for optimizing the variation of multiple injections on the engine performance and emissions of a DI diesel engine. The results show that, the combination of high pressure multiple injections with cooled EGR produces better overall results than the combination of low injection pressure multiple injections without EGR.

Jafarmadar and Zehni [11] studied the effect of split injection on combustion and pollution of a DI diesel engine by Computational Fluid Dynamics (CFD) code. The results show that 25% of total fuel injected in the second pulse, reduces the total soot and NOx emissions effectively in DI diesel engines. In addition, the optimum delay dwell between the two injection pulses was about 25ºCA.

Showry and Rajo [12] carried out the effect of triple injection on combustion and pollution of a DI diesel engine by FLUENT CFD code and concluded that 10° is an optimum delay between the injection pulses for triple injection strategy. The showed that split injections take care of reducing of PM without increasing of NOx level.

As mentioned, the effect of split injection on combustion and emission of DI Diesel engines has been widely studied up to now. However, for IDI diesel engines, the study of split injection strategy in order to reduce emissions is confined to the research of Iwazaki et al [13] that investigated the effects of early stage injection and two-stage injection on the combustion characteristics of an IDI diesel engine experimentally. The results indicated that NOx and smoke emissions are improved by two-stage injection when the amount of fuel in the first injection was small and the first injection timing was advanced from -80 to -100° TDC.

At the present work, the effect of the split injection on combustion and pollution of DI and IDI diesel engines is studied at full load state by the CFD code FIRE. The target is to obtain the optimum split injection case in which the total exhaust NOx and soot concentrations are more reduced than the other cases. Three different split injection schemes, in which 10, 20 and 25% of total fuel injected in the second pulse, have been considered. The delay dwell between injections is varied from 5°CA to 30°CA with the interval 5°CA.

2. Initial and boundry conditions

Calculations are carried out on the closed system from Intake Valve Closure (IVC) at 165°CA BTDC to Exhaust Valve Open (EVO) at 180°CA ATDC. Fig. 1-a and Fig. 1-b show the numerical grid, which is designed to model the geometry of combustion chamber of IDI

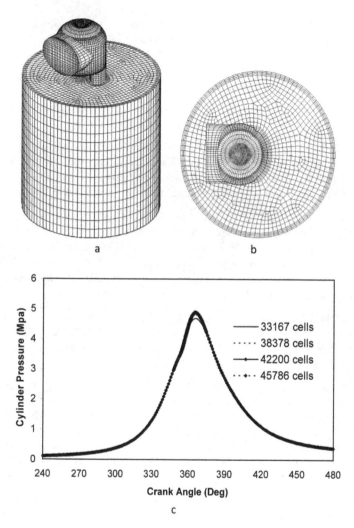

Figure 1. a. Mesh of the Lister 8.1 indirect injection diesel engine; b. Top view of the mesh; c. Grid dependency based on the in-cylinder pressure

engine and contains a maximum of 42200 cells at BTDC. As can be seen from the figure Fig. 1-c, Grid dependency is based on the in-cylinder pressure and present resolution is found to give adequately grid independent results. There is a single hole injector mounted, which is in pre-chamber as shown in fig. 2-a. In addition, details of the computational mesh used in DI are given in Fig. 2-b. The computation used a 90 degree sector mesh (the diesel injector has four Nozzle holes) with 25 nodes in the radial direction, 20 nodes in the azimuthal direction and 5 nodes in the squish region (the region between the top of the piston and the cylinder head)at top dead center. The ground of the bowl has been meshed with two continuous layers for a proper calculation of the heat transfer through the piston wall. The final mesh consists of a hexahedral dominated mesh. Number of cells in the mesh was about 64,000 and 36,000 at BDC and TDC, respectively. The present resolution is found to give adequately at DI engine. Initial pressure in the combustion chamber is set to 86 kPa and initial temperature is calculated to be 384K, and swirl ratio is assumed to be on quiescent condition. Boundary temperatures for head, piston and cylinder are 550K, 590K and 450K, respectively. Present work is studied at full load mode and the engine speed is 730 rpm. All boundary temperatures were assumed to be constant throughout the simulation, but allowed to vary with the combustion chamber surface regions.

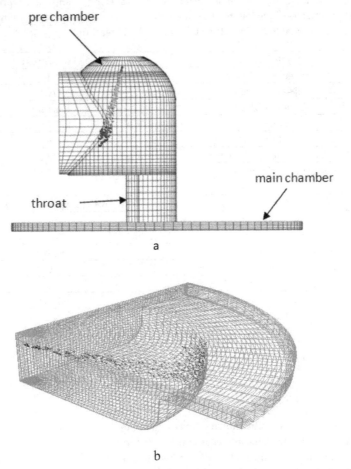

Figure 2. a. Spray and injector coordinate at pre-chamber; b Computational mesh with diesel spray drops at 350°CA, single injection case for DI engine.

3. Model formulation

The numerical model is carried out for Lister 8.1 indirect injection diesel engine and OM355 DI engine with the specification given on tables 1 and 2, repetivily. The governing equations for unsteady, compressible, turbulent reacting multi-component gas mixtures flow and thermal fields are solved from IVC to EVO by the commercial AVL-FIRE CFD code [14]. The turbulent flow within the combustion chamber is simulated using the RNG k-ε turbulence model, modified for variable-density engine flows [15].

The standard WAVE model, described in [16], is used for the primary and secondary atomization modeling of the resulting droplets. At this model, the growth of an initial perturbation on a liquid surface is linked to its wavelength and other physical and dynamical parameters of the injected fuel at the flow domain. Drop parcels are injected with characteristic size equal to the Nozzle exit diameter (blob injection).

Cycle Type	Four Stroke
Number of Cylinders	1
Injection Type	IDI
Cylinder Bore	114.1 mm
Stroke	139.7 mm
L/R	4
Displacement Volume	1.43 lit.
Compression Ratio	17.5 : 1
$V_{pre-chamber}/V_{TDC}$	0.7
Full Load Injected Mass	$6.4336e - 5$ kg per Cycle
Power at 850 rpm	5.9 kW
Power at 650 rpm	4.4 kW
Initial Injection Pressure	90 bar
Nozzle Diameter at Hole Center	0.003m
Number of Nuzzle Holes	1
Nozzle Outer diameter	0.0003m
Spray Cone Angle	10°
Valve Timing	IVO= 5° BTDC
	IVC= 15° ABDC
	EVO= 55° BBDC
	EVC= 15° ATDC

Table 1. Specifications of Lister 8.1 IDI diesel engine

The Dukowicz model is applied for treating the heat up and evaporation of the droplets, which is described in [17]. This model assumes a uniform droplet temperature. In addition, the droplet temperature change rate is determined by the heat balance, which states that the heat convection from the gas to the droplet either heats up the droplet or supplies heat for vaporization.

Piston shape	Cylindrical bore
No. of nozzles/injector	4
Nozzle opening pressure	195 (bar)
Cylinders	6, In-line-vertical
Bore * stroke	128 (mm) * 150 (mm)
Max. power	179 (kw) at 2200 (rpm)
Compression ratio	16.1:1
Max. torque	824 N m at 1400 (rpm)
Capacity	11.58 (lit)
IVC	61ºCA after BDC
EVO	60ºCA before BDC
Initial Injection Pressure	250(bar)

Table 2. Engine Specifications of OM-355 Diesel

A Stochastic dispersion model was employed to take the effect of interaction between the particles and the turbulent eddies into account by adding a fluctuating velocity to the mean gas velocity. This model assumes that the fluctuating velocity has a randomly Gaussian distribution [14].

The spray/wall interaction model used in this simulation was based on the spray/wall impingement model [18]. This model assumes that a droplet, which hits the wall was affected by rebound or reflection based on the Weber number.

The Shell auto-ignition model was used for modeling of the auto ignition [19]. In this generic mechanism, six generic species for hydrocarbon fuel, oxidizer, total radical pool, branching agent, intermediate species and products were involved. In addition, the important stages of auto ignition such as initiation, propagation, branching and termination were presented by generalized reactions, described in [14, 19].

The Eddy Break-up model (EBU) based on the turbulent mixing is used for modeling of the combustion in the combustion chamber [14]. This model assumes that in premixed turbulent flames, the reactants (fuel and oxygen) are contained in the same eddies and are separated from eddies containing hot combustion products. The rate of dissipation of these eddies determines the rate of combustion. In other words, chemical reaction occurs fast and the combustion is mixing controlled. NOx formation is modeled by the Zeldovich mechanism and Soot formation is modeled by Kennedy, Hiroyasu and Magnussen mechanism [20].

The injection rate profiles are rectangular type and consists of nineteen injection schemes, i.e. single injection and eighteen split injection cases(as shown in table 4). To simulate the split injection, the original single injection profile is divided into two injection pulses without altering the injection profile and magnitude. Fig. 3 illustrates the schematic scheme of the single and split injection strategy.

For the single injection case, the start of injection is at 348°CA and injection termination is at 387°CA. For all split injection cases, the injection timing of the first injection pulse is fixed at

348°CA. Three different split injection schemes, in which 10-20-25% of total fuel injected in the second pulse, has been considered. The delay dwell between injections is varied from 5°CA to 30°CA with the interval 5°CA.

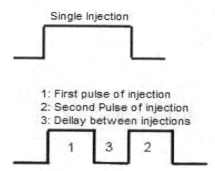

Figure 3. schematic scheme of single and split injection strategy.

4. Performance parameters

Indicated work per cycle is calculated from the cylinder pressure and piston displacement, as follows:

$$W = \int_{\theta_1}^{\theta_2} P dV \tag{1}$$

Where θ_1, θ_2 are the start and end of the valve-closed period, respectively (i.e. IVC= 15° ABDC and EVO= 55° BBDC). The indicated power per cylinder and indicated mean effective pressure are related to the indicated work per cycle by:

$$P(kW) = \frac{W(N.m)N(rpm)}{60000n} \tag{2}$$

$$IMEP = \frac{W}{V_d} \tag{3}$$

Where n=2 is the number of crank revolutions for each power stroke per cylinder, N is the engine speed in rpm and Vd is volume displacement of piston. The brake specific fuel consumption (BSFC) is defined as:

$$BSFC = \frac{\dot{m}_f}{P_b} \tag{4}$$

In Equation (1), the work is only integrated as part of the compression and expansion strokes; the pumping work has not been taken into account. Therefore, the power and

(ISFC) analyses can only be viewed as being qualitative rather than quantitative in this study.

5. Results and discussion IDI

Fig. 4 and Fig. 5 show the verification of computed and measured [21] mean in-cylinder pressure and heat release rate for the single injection case. They show that both computational and experimental data for cylinder pressure and heat release rate during the compression and expansion strokes are in good agreement.

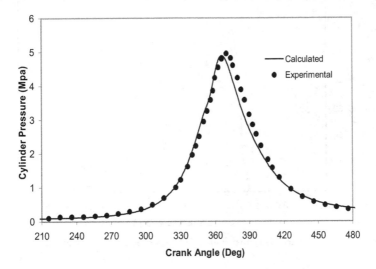

Figure 4. Comparison of calculated and measured [21] in-cylinder pressure, single injection case

The peak of cylinder Pressure is 4.88 Mpa, which occurs at 366°CA (4°CA after TDC). The start of heat release is at 351°CA for computed and measured results; in other words, the ignition delay dwell is 3°CA. It means that the ignition delay is quite close to the chemical ignition delay and that the physical ignition delay is very short, because of the rapid evaporation of the small droplets injected through the small injector gap at the start of injection. The heat release rate, which called "measured", is actually derived from the procured in-cylinder pressure data using a thermodynamic first law analysis as followed:

$$\frac{dq}{d\theta} = \frac{\gamma}{\gamma-1}p\frac{dV}{d\theta} + \frac{1}{\gamma-1}V\frac{dp}{d\theta} \qquad (5)$$

Where p and V are in-cylinder pressure and volume versus the crank angle θ, and γ=1.33.The main difference of computed and measured HRR is due to applying single zone model to combustion process with assuming γ=1.33 and observed at premixed combustion. The peak of computed HRR is 59J/deg that occurs at 364°CA, compared to the peak of measured HRR that is 53J/deg at 369°CA. The main validation is based on pressure in cylinder.

As a whole, the premixed combustion occurs with a steep slope and it can be one of the major sources of NOx formation.

Table 3 shows the variation of performance parameters for the single injection case, compared with the experimental data [21]. In contrast with the experimental results, it can be seen that model can predict the performance parameters with good accuracy.

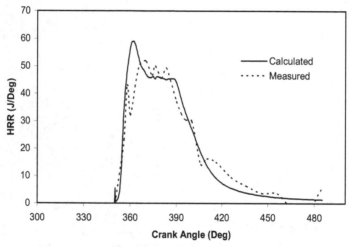

Figure 5. Comparison of calculated and measured [21] heat release rate, single injection case

parameters	Calculated	Experimental
Brake Power [kW]	4.53	4.65
BMEP [Bar]	5.33	5.47
Bsfc [g/kW.h]	310.72	302.7

Table 3. Comparison of calculated and measured [21] performance parameters, single injection case

Fig. 6 indicates that the predicted total in-cylinder NOx emission for the single injection case, agrees well with the engine-out measurements [21]. Heywood [22] explains that the critical time for the formation of oxides of nitrogen in compression ignition engines is between the start of combustion and the occurrence of peak cylinder pressure when the burned gas temperatures are the highest. The trend of calculated NOx formation in the prechamber and main chambers agrees well with the Heywood's explanations. As temperature cools due to volume expansion and mixing of hot gases with cooler burned gas, the equilibrium reactions are quenched in the swirl chamber and main chamber.

As can be seen from the Fig. 7, the predicted total in-cylinder soot emission for the single injection case agrees well with the engine-out measurements [21].

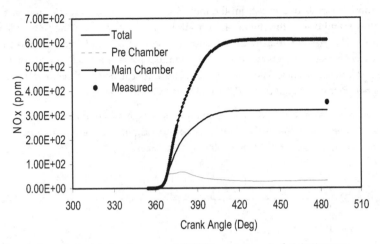

Figure 6. Comparison of calculated and measured [21] NOx emission, single injection case

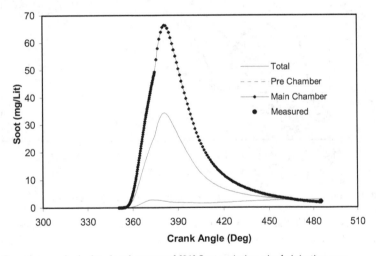

Figure 7. Comparison of calculated and measured [21] Soot emission, single injection case

Table 3 presents the exhaust NOx and soot emissions and performance parameters for the calculated single injection, and split injection cases. As can be seen, the lowest NOx and Soot emissions are related to the 75%-15-25% and 75%-20-25% cases respectively. In order to obtain the final optimum case, i.e. the case that involves the highest average of NOx and Soot reduction, a new dimensionless parameter is defined as:

The more of the total average emission reduction percentage results to the better optimum split injection case. Hence, it is concluded that the 75%-20-25% scheme with the average reduction percentage of 23.28 is the optimum injection case.

In addition, it can be deduced from the table 4 that in general, split injection sacrifices the brake power and Bsfc of the engine. This result is more apparent for the 80%-20% and 75%-25% cases. The reason is that with reduction of the first pulse of injection and increase of delay dwell between injections, premixed combustion as the main source of the power stroke is decreased. Hence, Bsfc is increased as well. As can be seen, the lowest brake power and highest Bsfc are related to the 75%-30-25% case.

Fig. 8 shows the NOx versus Soot emission for the single injection and split injection cases. As can be seen, the 75%-20-25% case is closer to origin, I, e. zero emission. Hence, this confirms that the case of 75%-20-25% is the optimum case.

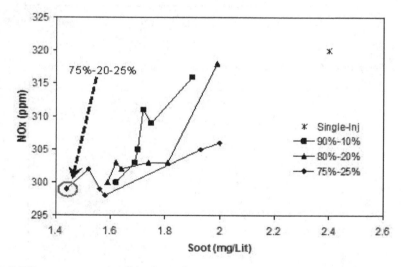

Figure 8. NOx versus soot emission for the single injection and split injection cases

It is of interest to notify that the optimum injection scheme for DI diesel engine at full load state is 75%-25-25% [11]. It means that the first and second injection pulses for the DI and IDI diesel engines are the same. The difference is related to the delay dwell between injections. I.e. the delay dwell for the optimum IDI split injection case is 5°CA lower than that of DI because of high turbulence intensity and fast combustion.

It is noticeable to compare the spray penetration, in-cylinder flow field, combustion and emission characteristics of the single injection and optimum injection cases to obtain valuable results.

The normalized injection profile versus crank angle for the single injection and 75%-20-25% cases is shown in the fig. 9. In this profile, actual injection rate values divided to maximum injection rate and this normalized injection rate profile is used by CFD code.

	NOx (ppm)	Soot (mg/lit)	NOx Reduction (%)	Soot Reduction (%)	Average reduction (%)	Pb (kW)	Bsfc (gr/kw.h)
Single Inj	320	2.4	0	0	0	4.53	310.72
90%-5-10%	316	1.9	1.25	20.83	11.04	4.51	312.1
90%-10-10%	309	1.75	3.43	27.08	15.25	4.4	319.9
90%-15-10%	300	1.62	6.25	32.5	19.37	4.32	325.8
90%-20-10%	311	1.72	2.81	28.33	15.57	4.37	322.1
90%-25-10%	303	1.69	5.31	29.58	17.44	4.33	325.08
90%-30-10%	305	1.7	4.68	29.16	16.92	4.35	323.58
80%-5-20%	318	1.99	0.6	17.08	8.84	4.36	322.84
80%-10-20%	303	1.81	5.31	24.58	14.94	4.31	326.58
80%-15-20%	303	1.74	5.31	27.5	16.4	4.2	335.14
80%-20-20%	302	1.64	5.62	31.66	18.64	4.14	340
80%-25-20%	303	1.62	5.31	32.5	18.9	4.13	340.82
80%-30-20%	300	1.59	6.25	33.75	20	4.12	341.65
75%-5-25%	306	2	4.37	20	12.18	4.29	328.11
75%-10-25%	305	1.93	4.68	19.58	12.13	4.18	336.74
75%-15-25%	**298**	1.58	6.87	34.16	20.51	4.12	341.65
75%-20-25%	299	**1.44**	6.56	40	**23.28**	4.04	348.41
75%-25-25%	302	1.52	5.62	36.66	21.14	4	351.9
75%-30-25%	299	1.56	6.56	35	20.78	3.97	354.55

Table 4. Exhaust emissions and performance parameters for the single injection and split injection cases

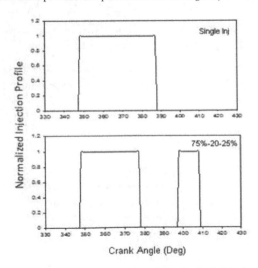

Figure 9. The normalized injection profile versus crank angle for the single injection and 75%-20-25% cases

Fig. 10-a and Fig. 10-b represent respectively front and top views of the evolution of the spray penetration and velocity field at various crank angles in horizontal planes of the pre and main combustion chambers and planes across the connecting throat for the single injection and 75%-20-25% cases. It can be seen that the maximum velocity in throat section are lower at all crank angles because of the large area this section than the other data in the literatures [23]. Generally, for all the cases, the maximum velocity of the flow field is observed at the tip of the spray, swirl chamber throat and some areas of the main chamber that is far from the cylinder wall and cylinder head.

As can be seen, at various crank angles, the main difference of the in-cylinder flow field between the single injection and split injection cases is due to the fuel injection scheme. In other words, the amount of the fuel spray and the crank position in which the spray is injected. The aerodynamic forces decelerate the droplets for the both cases. The drops at the spray tip experience the strongest drag force and are much more decelerated than droplets that follow in their wake.

At 370°CA, air entrainment into the fuel spray can be observed for the both cases. Hence, Droplet velocities are maximal at the spray axis and decrease in the radial direction due to interaction with the entrained gas.

Although the amount of the fuel spray for the single injection case is higher than the 75%-20-25% case at 370°CA, the flow field difference is not observed obviously. The more quantity of the fuel spray for the single injection case causes the maximum velocity of the single injection case to be higher by about 1m/s than the 75%-20-25% case.

At 380°CA, the fuel spray is cut off for the 75%-20-25% case. Since the entrained gas into to the spray region is reduced, the flow moves more freely from the swirl chamber to the main chamber. Hence, as can be seen from the front view, the maximum flow field velocity for the 75%-20-25% case is higher compared to the single injection case.

At 390°CA, from the top view, the distribution of the flow field in the whole of the main chamber is visible. In addition, some local swirls can be seen in the main chamber that are close to the swirl chamber. From the front view, it is observed that the gas coming from the pre-chamber reaches the opposite sides of cylinder wall. This leads to the formation of the two eddies occupying each one-half of the main chamber and staying centered with respect to the two half of the bowl.

The start of the second injection pulse for the 75%-20-25% case is observed At 400°CA. since the swirl intensity in the swirl chamber is reduced at this crank position, the interaction between the flow field and spray is decreased to somehow and flow in the pre-chamber is not strongly influenced by the fuel spray. Hence, the maximum velocity of the flow field for the 75%-20-25% case remains lower compared to the single injection case.

Fig. 11 indicates the history of heat release rate, cylinder pressure, temperature, O2 mass fraction, NOx and soot emissions for the single injection and 75%-20-25% cases.

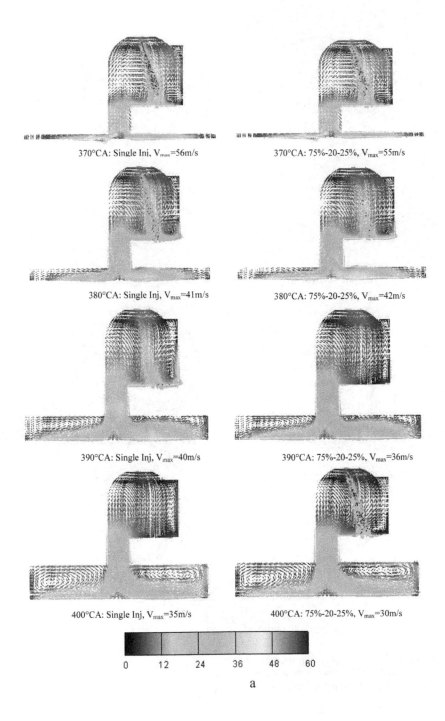

370°CA: Single Inj, V_{max}=56m/s

370°CA: 75%-20-25%, V_{max}=55m/s

380°CA: Single Inj, V_{max}=41m/s

380°CA: 75%-20-25%, V_{max}=42m/s

390°CA: Single Inj, V_{max}=40m/s

390°CA: 75%-20-25%, V_{max}=36m/s

400°CA: Single Inj, V_{max}=35m/s

400°CA: 75%-20-25%, V_{max}=30m/s

0   12   24   36   48   60

a

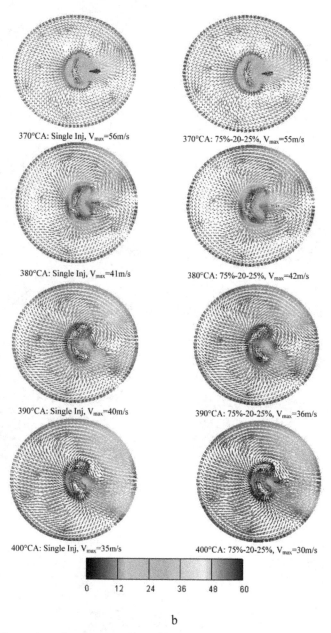

b

Figure 10. a. Comparison of spray penetration and velocity filed at various crank angles for the single injection and 75%-20-25% cases, front view; b. Comparison of spray penetration and velocity filed at various crank angles for the single injection and 75%-20-25% cases, top view.

Fig. 11-a shows that the amount of heat release rate for the both cases is to somehow equal until 380°CA. It is due to the fact that the first injection pulse for the 75%-20-25% case, lasts to 377°CA. Hence, the premixed combustion for the both cases does not differ visibly. For the 75%-20-25% case, The second peak of heat release rate occurs at 410°CA and indicates that a rapid diffusion burn is realized at the late combustion stage and it affects the in-cylinder pressure, temperature and soot oxidation.

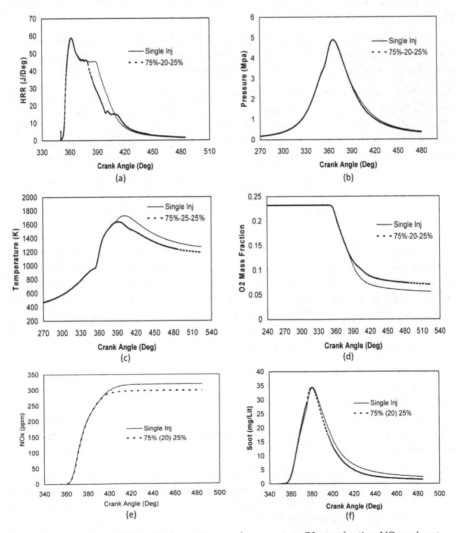

Figure 11. Comparison of HRR, cylinder pressure and temperature, O2 mass fraction, NOx and soot histories for the single injection and 75%-20-25% cases

Fig. 11-b shows that the second injection pulse of the 75%-20-25% does not cause to the second peak of the cylinder pressure. It is because the rate of decrease of cylinder pressure due to the expansion stroke is almost equal to the rate of increase of cylinder pressure due to the diffusion combustion.

Fig. 11-c compares temperature trend for 75%-20-25% and single injection cases. As can be seen, for the 75%-20-25% case, the peak of the temperature advances about 11°CA compared to the peak of the temperature of the single injection case. In addition, the temperature reduction reaches to 90°K.

Fig. 11-d presents the in cylinder O_2 mass fraction. It can be seen that after 385°CA, the O2 concentration for the 75%-20-25% case is higher than that for the single injection case. In other words, oxygen availability of 75%-20-25% case is better.

As Fig. 11-e and Fig. 11-f indicate, the reduction of the peak of the cylinder temperature for the 75%-20-25% case, causes to the lower NOx formation. In addition, soot oxidation for the 75%-20-25% case is higher compared to that for the single injection case. It is because for the 75%-20-25% case, the more availability of oxygen in the expansion stroke compensates the decrease of temperature due to the split injection. Hence, soot oxidation is increased.

Fig. 12-a and Fig. 12-b represent respectively front and top views of the contours of equivalence ratio, temperature, NOx and soot emissions at various crank angles in horizontal planes of the main combustion chamber and planes across the connecting throat for the single injection and 75%-20-25% cases.

At 400°CA, for the both cases, the majority of NOx is located in the two half of the main chamber, whilst soot is concentrated in the swirl chamber, throat section and main chamber. For the both cases, a local soot-NOx trade-off is evident in the swirl chamber, as the NOx and soot formation occur on opposite sides of the high temperature region. Equivalence ratio contour of the 75%-20-25% case confirms that injection termination and resumption, causes leaner combustion zones. Hence, the more reduction of soot for the 75%-20-25% is visible compared to the single injection case.

NOx and soot formation depend strongly on equivalence ratio and temperature. It is of interest to notice that the area which the equivalence ratio is close to 1 and the temperature is higher than 2000 K is the NOx formation area. In addition, the area which the equivalence ratio is higher than 3 and the temperature is approximately between 1600 K and 2000 K is the Soot formation area. The area from 1500 K and equivalence ratio near to 1 is defined as soot oxidation area [24, 25].

At 410°CA, due to the second injection pulse, increase of the equivalence ratio is observed in the swirl chamber of the 75%-20-25% case. Hence, the enhancement of the soot concentration close to the wall spray impingement is observed. For the other areas, soot oxidation for the 75%-20-25% case is higher compared to the single injection case. In addition, NOx reduction tendency is visible in the right side of the main chamber of the 75%-20-25% case compared to the single injection case. It is because more temperature reduction occurs for the 75%-20-

25% case due to the lower premixed combustion. The diffusion combustion of the second injection pulse does not affect NOx formation significantly.

The final form of the distribution of equivalence ratio, temperature, NOx and soot contours can be seen at 420°CA. The reduction of soot and NOx for the   75%-20-25% case is clear compared to the single injection case. This trend is preserved for the both cases until EVO.

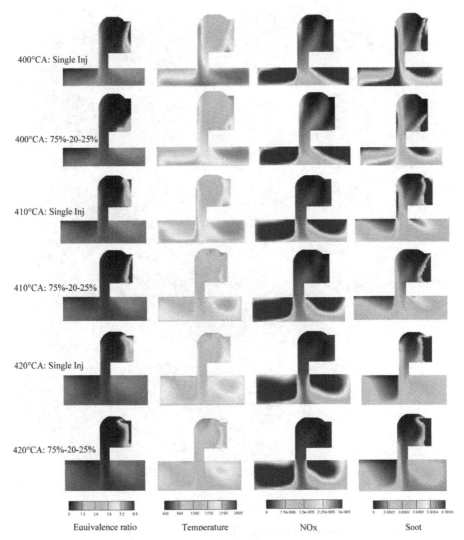

a

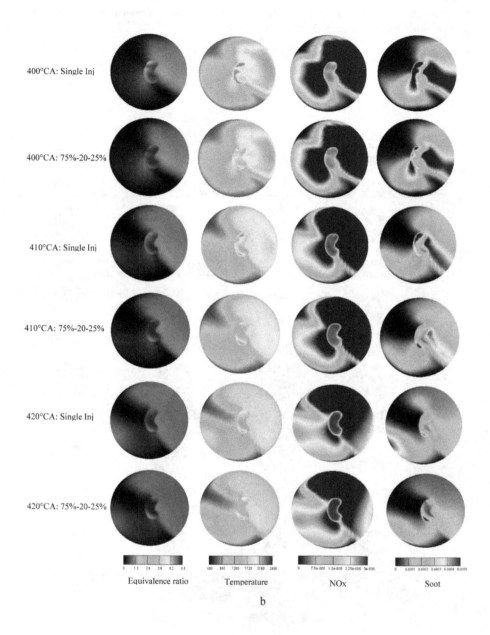

Figure 12. a. Contours of equivalence ratio, temperature, NOx and Soot at different crank angles, front view b. Contours of equivalence ratio, Temperature, NOx and Soot at different crank angles, top view

6. Results and discussion of DI diesel engine

Fig. 13 shows the cylinder pressure and the rate of heat release for the single injection case. As can be seen from HRR curve, the peak of the heat release rate occurs at 358ºCA (2ºCA before TDC). The premixed combustion occurs with a steep slope and it can be one of the major sources of NOx formation. The good agreement of predicted in-cylinder pressure with the experimental data [27] can be observed.

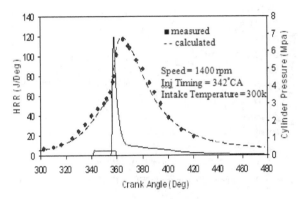

Figure 13. HRR and Comparison of calculated and measured [27] in-cylinder pressure, single njection case.

Fig. 14 and Fig. 15 imply that the predicted total in-cylinder NOx and soot emissions for the single injection case, agree well with the engine-out measurements [27].

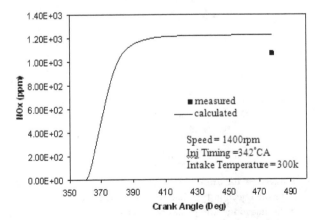

Figure 14. Comparison of calculated and measured [27] NOx emission, single injection case.

Fig. 16 shows the trade-off between NOx and soot emissions at EVO when the injection timing is varied. As indicated, the general trends of reduction in NOx and increase in soot when injection timing is retarded can be observed and it is independent on injection strategy. The reason is that it causes the time residence and ignition delay to be shorter, resulting in a less intense premixed burn and soot formation increases; in addition, the less temperature in different parts of combustion chamber keeps the soot oxidation low but decreases the formation of thermal NOx.

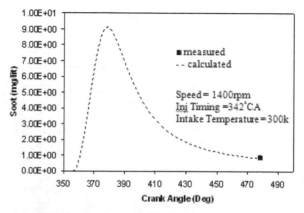

Figure 15. Comparison of calculated and measured [27] soot emission, single injection case.

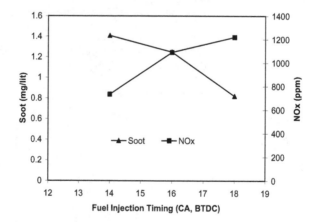

Figure 16. The effect of injection timing on NOx and Soot trade-off, single injection case.

To simulate the split injection, the original single injection profiles are split into two injection pulses without altering the injection profile and magnitude. In order to obtain the optimum dwell time between the injections, three different schemes including 10%, 20% and 25% of the total fuel injected in the second pulse are considered.

Fig. 17 shows the effect of delay dwell between injection pulses on soot and NOx emissions for the three split injection cases. For all the cases, the injection timing of the first injection pulse is fixed at 342ºCA.

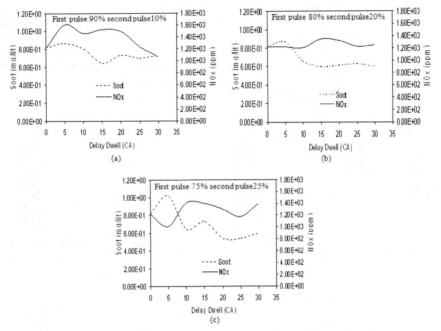

Figure 17. The variation of soot and NOx at different delay dwells, split injection cases.

The variation trend of curves in Fig. 17 is very similar to the numerical results obtained by Li et al [28]. As can be seen, the optimum delay dwell between the injection pulses for reducing soot with low NOx emissions is about 25°CA. The evidences of Li, J. et al. [28], which has used a phenomenological combustion model, support this conclusion.   Table 5 compares Exhaust NOx and soot emissions for the single injection and optimum split injection cases for the optimum delay dwell. As shown, for the 75% (25) 25% case, NOx and soot emissions are lower than the other cases. It is due to the fact that the premixed combustion which is the main source of the NOx formation is relatively low. The more quantity of the second injection pulse into the lean and hot combustion zones, causing the newly injected fuel to burn rapidly and efficiently at high temperatures resulting in high soot oxidation rates. In addition, the heat released by the second injection pulse is not sufficient to increase the NOx emissions. Fig. 18 and Fig. 19 confirm the explanations.

Case	NOx (ppm)	Soot (mg/lit)
Single inj	1220	0.82
Split inj- 90% (25) 10%	1250	0.697
Split inj- 80% (25) 20%	1230	0.614
Split inj- 75% (25) 25%	1180	0.541

Table 5. Comparison of NOx and soot emissions among the single injection and optimum split injection cases for the optimum delay dwell (25°CA)

Fig. 18 shows the cylinder pressure and heat release rates for the three split injection cases in the optimum delay dwell i.e. 25°CA. As shown, split injection reduces the amount of premixed burn compared to the single injection case in Fig. 13. The second peak which appears in heat release rate curves of split injection cases indicates that a rapid diffusion burn is realized at the late combustion stage and it affects the in-cylinder pressure and temperature. The calculated results of the cylinder pressure and HRR for the optimum split injection cases show very good similarity with the results of the reference[28].

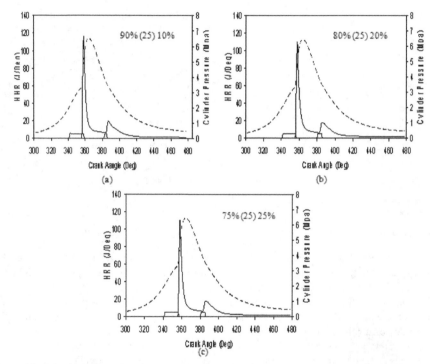

Figure 18. Cylinder HRR and pressure curves, optimum split injection cases.

Fig. 19 indicates the cylinder temperature for the single injection and optimum split injection cases. For the split injection cases, the two peaks due to the first and second injection pulses in contrast with the one peak in the single injection case can be observed. As shown, for the 75% (25) 25% case, the first and second peaks which are related to the premixed combustion and NOx formation are lower than the other cases. In addition, after the second peak, the cylinder temperature tends to increase more in comparison with the other cases and causes more soot oxidation. Hence, for the 75% (25) 25% case, NOx and soot emissions are lower than the other optimum cases.

Figure.20. shows the isothermal contour plots at different crank angle degrees for the 75% (25) 25% case at a cross-section just above the piston bowl. The rapid increase in temperature due to the stoichiometric combustion can be observed at 355ºCA. At 360ºCA, the injection termination can be observed which the in-cylinder temperature tends to be maximum. At 370ºCA, the fuel injection has been cut off and the cylinder temperature tends to become lower. As described, injection termination and resumption prevents not only fuel rich combustion zones but also causes more complete combustion due to better air utilization. The resumption of the injection can be observed at 380ºCA which causes the diffusion combustion and increases the temperature in the cylinder. At 390ºCA, the increase of cylinder temperature creates bony shape contours and soot oxidation increases.

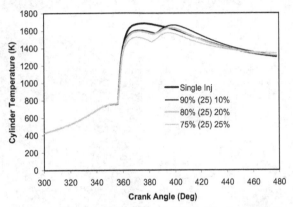

Figure 19. Comparison of cylinder temperature among the single injection and optimum split injection cases.

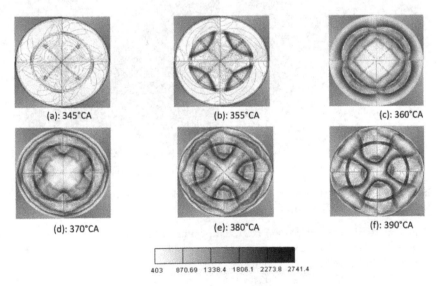

Figure 20. Isothermal contour plots of 75% (25) 25% case at different crank angle degrees.

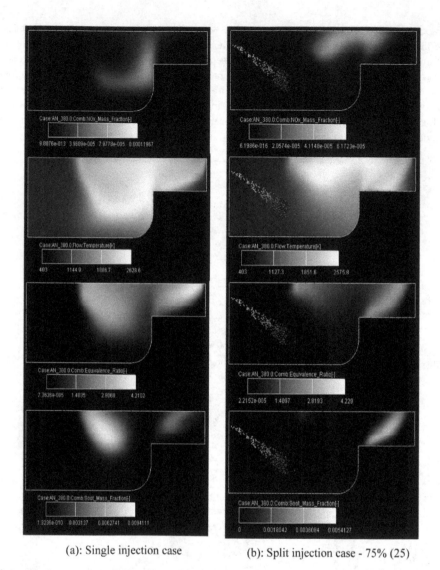

(a): Single injection case   (b): Split injection case - 75% (25)

Figure 21. Contour plots of NOx, temperature, equivalence ratio, and soot with fuel droplets at 380°CA for the single injection and 75% (25) 25% cases.

Fig. 21 compares the contour plots of NOx, temperature, equivalence ratio and soot for the single injection and 75% (25) 25% cases in a plane through the center of the spray at 380ºCA. As explained above, 380ºCA corresponds to a time when the second injection pulse has just started for the 75% (25) 25% case. For the two described cases, it can be seen that the area which the equivalence ratio is close to 1 and the temperature is higher than 2000 K is the

NOx formation area. In addition, the area which the equivalence ratio is higher than 3 and the temperature is approximately between 1600 K and 2000 K is the soot formation area. A local soot-NOx trade-off is evident in these contour plots, as the NOx formation and soot formation occur on opposite sides of the high temperature region. It can be seen that for the 75% (25) 25% case, NOx and soot mass fractions are lower in comparison with the single injection case. Because of the optimum delay dwell, the second injection pulse, maintains the low NOx and soot emissions until EVO.

7. Conclusion

At the present work, the effect of the split injection on combustion and pollution of DI and IDI diesel engines was studied at full load state by the CFD code. The target was to obtain the optimum split injection cases for these engine in which the total exhaust NOx and soot concentrations are the lowest.

Three different split injection schemes, in which 10, 20 and 25% of total fuel injected in the second pulse, was considered. The delay dwell between injections pulses is varied from 5°CA to 30°CA with the interval 5°CA.The results for IDI are as followed:

1. The calculated combustion and performance parameters, exhaust NOx and soot emissions for the single injection case showed a good agreement with the corresponding experimental data.
2. The lowest NOx and Soot emissions are related to the 75%-15-25% and 75%-20-25% cases respectively. Finally, optimum case was 75%-20-25% regarding the highest average of NOx and soot reduction.
3. The lowest brake power and highest BSFC quantity is due to the 75%-30-25% case.
4. Because in the literature, the optimum split injection scheme for DI diesel engines at full state was defined as 75%-25-25%, it was concluded that the difference of the optimum split injection scheme for DI and IDI diesel engines was related to the delay dwell between the injections.
5. The main difference of the in-cylinder flow field between the single injection and split injection cases is due to the fuel injection scheme.
6. For the 75%-20-25% case, the more availability of oxygen in the expansion stroke compensates the decrease of temperature due to the split injection. Hence, soot oxidation is increased.
7. The final form of the NOx and soot at 420°CA which preserve it's trend until EVO, confirms the more reduction of NOx and soot for the 75%-20-25% case in comparison with the single injection case.

In addition, the results for DI engins are as follow:

1. A good agreement of predicted in-cylinder pressure and exhaust NOx and soot emissions with the experimental data can be observed.
2. Advancing or retarding the injection timing can not decrease the soot and NOx trade-off by itself. Hence split injection is needed.

3. The optimum delay dwell between the injection pulses for reducing soot with low NOx emissions is about 25°CA. The results of phenomenological combustion models in the literature support this conclusion.
4. The calculated results of the cylinder pressure and heat release rate for the optimum split injection cases show very good similarity with the numerical results obtained by phenomenological combustion models.
5. For the 75% (25) 25% case, NOx and soot emissions are lower than the other cases. It is due to the fact that the premixed combustion which is the main source of the NOx formation is relatively low. The more quantity of the second injection into the lean and hot combustion zones, causing high soot oxidation rates. In addition, the heat released by the second injection pulse is not sufficient to increase the NOx emissions.
6. Contour plots of NOx, temperature, equivalence ratio and soot for the single injection and 75% (25) 25% cases at 380°CA show that the area which the equivalence ratio is close to 1 and the temperature is higher than 2000 K is the NOx formation area. In addition, the area which the equivalence ratio is higher than 3 and the temperature is approximately between 1600 K and 2000 K is the Soot formation area.

Abbreviations

BTDC   before top dead center
ATDC   after top dead center
EVO   exhaust valve opening
IDI   indirect injection engine
CA   crank angle
SOC   start of combustion
HRR   heat release rate
ID   ignition delay
BSFC   brake specific fuel consumption
ABDC   after bottom dead center
BBDC   before bottom dead center
BMEP   brake mean effective pressure

Author details

S. Jafarmadar
Mechanical Engineering Department, Technical Education Faculty, Urmia University, Urmia, West Azerbaijan, Iran

8. References

[1] Gomaa M, Alimin AJ, Kamarudin KA. Trade-off between NOx, soot and EGR rates for an IDI diesel engine fuelled with JB5. World Academy of Science, Engineering and Technology 2010; 62:449-450.

[2] Bianchi GM, Peloni P, Corcione FE, Lupino F. Numerical analysis of passenger car HSDI diesel engines with the 2nd generation of common rail injection systems: The effect of multiple injections on emissions. SAE Paper NO. 2001-01-1068; 2001.

[3] Seshasai Srinivasan, Franz X. Tanner, Jan Macek and Milos Polacek. Computational Optimization of Split Injections and EGR in a Diesel Engine Using an Adaptive Gradient-Based Algorithm. SAE paper, 2006-01-0059

[4] Shayler Pj, Ng HK. Simulation studies of the effect of fuel injection pattern on NOx and soot formation in diesel engines. SAE Paper NO. 2004-01-0116; 2004.

[5] Chryssakis CA, Assanis DN, Kook S, Bae C. Effect of multiple injections on fuel-air mixing and soot formation in diesel combustion using direct flame visualization and CFD techniques. Spring Technical Conference, ASME NO. ICES2005-1016; 2005.

[6] Lechner GA, Jacobs TJ, Chryssakis CA, Assanis DN, Siewert RM. Evaluation of a narrow spray cone angle, advanced injection strategy to achieve partially premixed compression ignition combustion in a diesel engine. SAE Paper No. 2005-01-0167; 2005.

[7] Ehleskog R. Experimental and numerical investigation of split injections at low load in an HDDI diesel engine equipped with a piezo injector. SAE Paper NO. 2006-01-3433; 2006.

[8] Sun YD, Reitz R. Modeling diesel engine NOx and soot reduction with optimized two-stage combustion. SAE Paper No. 2006-01-0027; 2006.

[9] Verbiezen K. et al. Diesel combustion: in-cylinder NO concentrations in relation to injection timing. Combustion and Flame, 2007; 151:333–346.

[10] Abdullah N, Tsolakis A, Rounce P, Wyszinsky M, Xu H, Mamat R. Effect of injection pressure with split injection in a V6 diesel engine. SAE Paper NO. 2009-24-0049; 2009.

[11] Jafarmadar S, Zehni A. Multi-dimensional modeling of the effects of split injection scheme on combustion and emissions of direct-injection diesel engines at full load state. IJE, 2009; Vol. 22, No. 4.

[12] Showry K, Raju A. Multi-dimensional modeling and simulation of diesel engine combustion using multi-pulse injections by CFD. International Journal of Dynamics of Fluids 2010; ISSN 0973-1784 Vol 6, NO 2, p. 237–248.

[13] Iwazaki K, Amagai K, Arai M. Improvement of fuel economy of an indirect (IDI) diesel engine with two-stage injection. Energy Conversion and Management, 2005; 30:447-459.

[14] AVL FIRE user manual V. 8.5, 2006.

[15] Han Z, Reitz R D.Turbulence modeling of internal combustion engines using RNG K-ε models. Combustion Science and Technology 1995; 106, p. 267-295.

[16] Liu AB, Reitz RD. Modeling the effects of drop drag and break-up on fuel sprays. SAE Paper NO. 930072; 1993.

[17] Dukowicz JK. Quasi-steady droplet change in the presence of convection. Informal report Los Alamos Scientific Laboratory. LA7997-MS.

[18] Naber JD, Reitz RD. Modeling engine spray/wall impingement. SAE Paper NO. 880107; 1988.

[19] Halstead M, Kirsch L, Quinn C. The Auto ignition of hydrocarbon fueled at high temperatures and pressures - fitting of a mathematical model. Combustion Flame 1977; 30:45-60.

[20] Patterson MA, Kong SC, Hampson GJ, Reitz RD. Modeling the effects of fuel injection characteristics on diesel engine soot and NOx emissions. SAE Paper NO. 940523; 1994.

[21] Khoushbakhti Saray R, Mohammahi Kusha A, Pirozpanah V. A new strategy for reduction of emissions and enhancement of performance characteristics of dual fuel engine at part loads. Int. J. Engineering 2010; 23:87-104.

[22] Heywood JB. Internal combustion engine fundamental. New York: McGraw Hill Book Company; 1988, p. 568-586.

[23] Sbastian S, Wolfgang S. Combustion in a swirl chamber diesel engine simulation by computation of fluid dynamics. SAE Paper NO. 950280; 1995.

[24] Yoshihiro H, Kiyomi N, Minaji I. Combustion improvement for reducing exhaust emissions in IDI diesel engine. SAE Paper NO. 980503; 1998.

[25] Hajireza S, Regner G, Christie A, Egert M, Mittermaier, H. Application of CFD modeling in combustion bowl assessment of diesel engines using DoE methodology. SAE Paper NO. 2006-01-3330; 2006.

[26] Carsten B. Mixture formation in internal combustion engines. Berlin: Springer publications; 2006, p. 226-231.

[27] Pirouzpanah, V. and Kashani, B. O., "Prediction of major pollutants emission in direct-injection dual-fuel diesel and natural gas engines", SAE Paper, NO. 1999-01-0841, (1999).

[28] Li, J., Chae, J., Lee, S. and Jeong, J. S., "Modeling the effects of split injection scheme on soot and NOx emissions of direct injection diesel engines by a phenomenological combustion model", SAE Paper, NO. 962062, (1996).

Chapter 2

Study of PM Removal Through Silent Discharge Type of Electric DPF Without Precious Metal Under the Condition of Room Temperature and Atmospheric Pressure

Minoru Chuubachi and Takeshi Nagasawa

Additional information is available at the end of the chapter

1. Introduction

For CO_2 reduction and prevention of global warming, clean diesel engine vehicles have been appeared in the market. These vehicles have already cleared the severe emission regulation, such as a new post long term regulation in Japan and also due to the good combustion efficiency, their fuel efficiency are better than gasoline engine vehicle. The latest these diesel engine vehicles equip DPF (Diesel Particulate Filter) sampling PM (Particulate Matter included in the exhaust gas) and DOC (Diesel Oxidation Catalyst) which is set upstream of DPF and bore much precious metals such as Pt. NO in exhaust gas is converted to NO_2 by DOC, and a system removing PM in oxygenation of the NO_2 is put to practical use [1][2]. However, generally the exhaust gas temperature is less than 250 deg. C in driving condition in the city areas, so, PM deposits in DPF without being oxidized and removed and then it may invite aggravation of the fuel efficiency from the increase of the exhaust pressure loss and an engine output power decline. For increasing exhaust gas temperature more than 250 deg. C, it is general that put together with a system increasing amount of fuel injection of the engine [3], but the fuel efficiency turns worse. In addition, the increase of the pressure loss may influence on power or in the worst case, the durability of the DPF such as melting down [4] [5]. Here, there are some problems that above exhaust system need much precious metal to the DOC or DPF, the fuel efficiency turns worse, the PM consecutive removal at the low temperature is not possible. Against these problems, in this study, the Silent discharge type of electric DPF (SDeDPF) was devised originally instead of (DOC + DPF + post fuel Injection) system with unique electrode material of MFS (Metal Fiber Sheet) and Turbulent Block which makes lower speed of an exhaust gas flow. The aim of this SDeDPF is PM removal without using precious metal under a room temperature and an atmospheric pressure condition continually [6][7]. Finally, it contributes improving fuel consumption and CO_2 reduction of the diesel engine vehicle.

2. Basic structure of SDeDPF

2.1. What is silent discharge?

The basic structure of silent discharge is shown in Fig. 1. This serves as an electric discharge cell which is a base of this research SDeDPF. The dielectrics called a barrier between the MFS (Metal Fiber Sheet) electrodes which face is inserted. Furthermore, between the dielectric and MFS electrode, the electric discharge gap (Gap d) by a spacer is prepared. If high frequency (several kilohertz) and the high voltage (several kilovolts) are added between electrodes, this whole space will be dyed a blue-purple color almost uniformly and countless minute electric discharge (micro streamer electric discharge) occurs simultaneously.

Although this electric discharge is dielectric barrier discharge or an electric discharge type only called barrier discharge, existence of a dielectric barrier has stopped progress to spark electric discharge or arc discharge. For the reason, it is also called silent discharge especially from it being the quiet electric discharge which does not emit the crashing sound at the time of electric discharge.

Since the continuation time of minute electric discharge is as short as ns (nanometer second) order, the energy transmission from an electron to ion and a gas molecule can be disregarded, and ion and a gas molecule are still room temperature.

Although the dielectric barrier discharge under atmospheric pressure condition is low-temperature (gas temperature) operation, electron temperature is high. So since it excels in the generative capacity of activated species (radical; the atom and molecule which were rich in chemical reactivity) it is applied to an ozone generator, the surface treatment of a polymer material, disassembly of the environmental pollutant, etc. in many fields [8]. Moreover, current does not flow directly between electrodes like arc discharge or corona discharge through a barrier between electrodes, and it is the big feature that there is also little power consumption.

PM removal of SDeDPF of this research applied oxidation reaction, C (the carbon C (subject of PM) to CO_2 by the activated oxygen sorts (O, OH, O_3, NO_2) generated in the electric discharge space. The component parts adopted as the electric discharge cell of SDeDPF of this research in Fig. 1 are explained concretely. The dielectrics used as a barrier is a ceramic board (Al_2O_3) of square of 114 mm generally marketed, t = 1 mm in thickness, and dielectric constant ε = 8.5. And in order to form the gap of electric discharge space, the 1mm thickness glass board was set up between the dielectric and the electric discharge electrode at the spacer. MFS (Metal Fiber Sheet,; 80% porosity with SUS material and the sintering sheet of 30 micrometers of fiber diameter) was adopted to an electrode, and it is one of the features of this research.

2.2. Structure of DPF System

2.2.1. Structure of current DPF system

The latest current DPF system is shown in Fig. 2 [1] [2]. Oxidation catalyst converter (DOC) which is a Channel-Flow type is arranged at the upper stream of DPF. DOC is supported by the precious metal catalyst of platinum (Pt) etc. Then NO in exhaust gas is converted into NO_2 at a catalyst reaction by precious metal. After DOC, DPF is arranged and when exhaust gas passes a porous ceramic wall of DPF of the Wall-Flow type, and then PM was trapped on the wall. PM (Carbon subject) trapped by DPF was oxidized to CO_2 by NO_2 generated by DOC and then it was exhausted as a clean gas. This system has already been put in practical use as DPF of a continuation reproduction system (ex. CRT; Continuously Regenerating Trap etc.)[2]. However, it is necessary to maintain catalyst temperature at not less than 250 deg. C for a catalyst reaction. By less than it, DOC does not work well as a catalyst so that PM trapped by DPF is deposited in DPF without carrying out oxidization removal. Then pressure in DPF rises and there is a possibility of causing the durability of DPF and also the fall of engine power.

For the improvement, responding to a pressure increase in DPF, the precise fuel injection control which makes engine fuel injection amount and times increase, raises exhaust gas temperature, and promotes a catalyst reaction is needed.

2.2.2. The preceding example of DPF applied electric discharge

Moreover, instead of the above-mentioned CRT, many methods of PM removal by electric discharge are also proposed [9]. The latest example of a proposal that transposed DPF made from ceramics to electric discharge type of PM removal equipment is shown in Fig. 3 and Fig. 4 [10] DOC is arranged like CRT in the upper stream of exhaust gas flow and the electric discharge equipment is arranged in the latter part. As Fig. 4 upper drawing, the electrode (Electrode) of metallic foil is printed on the surface of a dielectric (Dielectrics), and wavy-mesh Electrode which becomes a more pair of electrodes is set up on it and then electric discharge is generated between both these electrodes. And Fig. 4 below drawing, an exhaust gas passage is formed by laminating. It is a system which makes this space generate electric discharge and from which it removes PM. In this electric discharge structure, electric discharge becomes the streamer electric discharge into which current flows directly between both electrodes. Although the electrode is spread around the large field, electric discharge concentrates on the shortest distance part between electrodes. So it is hard to change with uniform electric discharge in the whole field and space of an exhaust passage. For the reason, generation of OH radical and activated oxygen sorts etc. are restricted to a part of space, so it is difficult to improve the removal performance of PM. Therefore, it seems that DOC is arranged in the upper stream of the exhaust gas flow. However, as same as the explanation of the above-mentioned CRT, since DOC cannot perform NO_2 generation by a catalyst reaction below at catalyst conversion temperature around 250 deg. C., it cannot perform removal of PM. Moreover, since it is electric discharge between both electrodes directly, it is thought that there are some concerns of that power consumption becomes increase due to be easy to flow greatly in current, generating of the crashing sound at the time of electric discharge, high temperature degradation of electrodes etc..

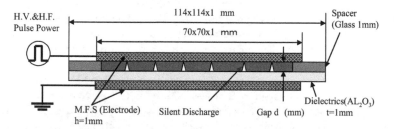

Figure 1. Basic cell of Silent electric Discharge

Figure 2. Current DPF system of continuously regenerating trap [1][2]

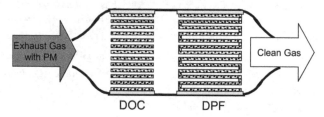

Figure 3. Example of plasma reactor system

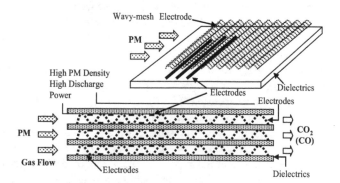

Figure 4. Example of structure of plasma reactor [10]

2.2.3. Structure of silent discharge type of electric DPF (SDeDPF) of this research

The structure of the SDeDPF final specification (Ver. 6) of this research is shown in Fig. 5 and 6. First, the front view seen from the exhaust upper stream side is shown in Fig. 5. It is the big feature that it is the Channel-Flow type with which exhaust gas passes through the electric discharge space ([7] Discharging Space) in a figure, and so that there is little pressure loss. Oxidization removal of the PM (Carbon) which passes through electric discharge space is efficiently carried out in the whole electric discharge space by OH radical and activated oxygen sorts (O, O_3) and NO_2 etc. which occurs by silent electric discharge. Since the dielectric ([1] Dielectric) is set between electrodes, large current does not flow directly so that there is little power consumption, there is little generating of the noisy sound at the time of electric discharge, as for the feature, there are also few rises in heat of an electrode part. Moreover, silent electric discharge differs from the electric discharge type example of PM removal equipment of above-mentioned Fig. 3 and 4, the generative capacity of OH radical and activated Oxygen sorts (O, O_3) and NO_2 etc. are high also under room temperature and atmospheric pressure conditions, without completely using the catalyst precious metals from the knowledge from other articles [11] [12] [13]. It is expectable to consider it as efficient continuous PM removal equipment.

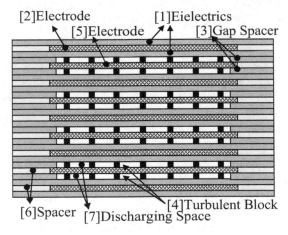

Figure 5. SDeDPF (front view)

Fig. 6 is the plan view (above) which looked at the part of the electric discharge space from the exhaust upper stream, and its AA sectional view (below). As shown in AA sectional view (below), it based the basic structure of the silent electric discharge of Fig. 1. A dielectric ([1] Dielectric), the high-voltage side electrode ([5] MFS) and the grounding side electrode ([2] MFS) form electric discharge space ([7] Discharging Space, Gap d = 1 mm). The high-voltage side electrode ([5] MFS) of the central part constitutes the primitive cell (Basic Cell) is shared in two electric discharge space. Furthermore four steps of this were stacked, finally SDeDPF is constituted with eight layers of electric discharge space as shown in Fig. 5.

(a) Plan view

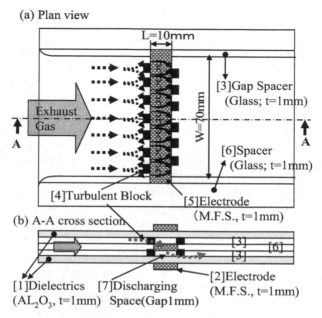

Figure 6. SDeDPF (plan view and A-A cross section view)

AS mentioned above, one of the features of SDeDPF is adoption of MFS (the sintered sheet of SUS material fiber, fiber diameter of 30 μm, 1 mm in thickness, and 80% porosity) as the electrode. The aim of the adoption of MFS is utilizing the characteristics, such as reduction of the pressure loss by 80% porosity in electric discharge space, and easy excitation of the random micro streamer electric discharge by the rough surface of a sheet, and outstanding high temperature oxidation resistance.

2nd feature is adoption of the turbulent blocks ([4] Turbulent Block: TB). As shown in Fig.6, seven TBs are set at the upper part of an electrode and six TBs are set at the lower part of an electrode and these TBs pick up the high voltage side electrode from the upper and lower sides of it in an electric discharge space part ([7]). Aims of an adoption of these TBs are following, prevention of a deformation and a position gap and keeping stabilization of electric discharge. Moreover, finally there is an aim to improve PM removal ratio more by turbulent-flowing and a speed fall of an exhaust gas with TBs and prolonging electric discharge irradiation time to PM.

2.3. Mechanism of PM removal

As shown in Fig. 7, if plasma irradiated in an exhaust gas, the electron of the high energy which arises to electric discharge space in the state of high-voltage electric discharge will collide with the molecule of the oxygen contained in an exhaust gas, water, and nitrogen oxide. A mechanism as shown in Table 1, reactive oxygen species (O, O3), OH radical and N radical in NO2 with high oxidation reaction nature are generated [11]. And then PM (main ingredients C Solid) in an exhaust gas will be oxidized and removed as CO2 by these reactive oxygen sorts and radicals [12]. Although N2 in the air has already oxidized to NO in the combustion process in the actual engine, in plasma discharge, N2 in the atmosphere is converted into NO2 of strong oxide and accelerates PM oxidization removal [13].

Generally, compared with the conversion from NO to NO2 and the O3 generation by barrier discharge, the generation energy of O3 is far small than it of NO2. (i.e. O3 is easy to be generated far.) For generation of NO2, like a chemical reaction formula of N (NO, NO2) Radical Generation in Table 1, there is a report [14] that O and O3 are involving greatly for generation of NO2. And like a chemical reaction formula of Activated Oxygen (O, O3) Generation in Table 1, since O3 is generated by work of O, O considers again that it must be dominant in the place near room temperature. Furthermore, there is a report [15] that energy required for the electron which arose by electric discharge to trigger the dissociative reaction of a gas molecule (O2, N2, H2O), that it is 3 to 4 times as required as O2 and H2O in the case of N2 and N radical generation speed is slow. From the generation equation of O and OH like as Table 1, O and OH were thought that it is dominant in the place near room temperature.

Like Table 1, an oxidation catalyst converts N2 and NO in the exhaust gas into NO2 at a catalyst reaction, and PM removal of the present diesel engine vehicle equipped with the oxidation catalyst and DPF is carrying out oxidization removal of the PM trapped in DPF by this NO2.

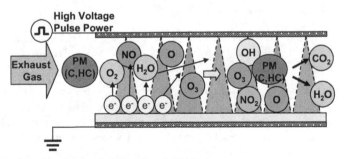

Figure 7. PM removal mechanism by Silent electric Discharge

Activated Oxygen (O, O_3)Generation	OH Radical Generation
$O_2 + e \rightarrow O^- + O + e$ $O + O_2 + M \rightarrow O_3 + M^*$	$H_2O + e \rightarrow OH + H + e$ $O + H_2O \rightarrow 2OH$
N (NO,NO_2)Radical Generation	PM(C solid) Oxidation
$N2 + e \rightarrow N + N + e$ $N + O_2 \rightarrow NO + O$ $O_3 + NO \rightarrow NO_2 + O_2$	$C + 2O \rightarrow CO_2$ $C + 2OH \rightarrow CO_2 + 2H$ $C + 2NO_2 \rightarrow CO_2 + 2NO$

Table 1. PM (C Solid) removal mechanism by Silent electric Discharge

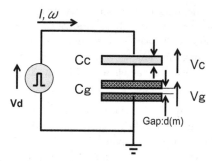

Figure 8. Real circuit of Silent electric Discharge

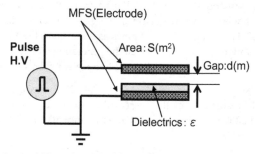

Figure 9. Equivalent circuit of Silent electric Discharge

Although for this catalyst reaction, the atmosphere temperature of not less than 250 deg. C is required, in silent electric discharge as mentioned above, oxidation reaction of PM by O, OH or O_3, and NO_2 occurs at even near room temperature.

Therefore, SDeDPF is in predominance about the point which can carry out oxidization removal to the ability of continuous PM removal in the state of room temperature against that the combination of conventional (DOC+DPF) can not perform PM removal in the state of room temperature. In order that the conventional PM removal system may solve this state, the present condition is increasing suitably the amount of engine fuel injection, raising exhaust gas temperature to not less than 250 deg. C, activating a catalyst, and promoting generation of NO_2.

The generation process of O, O_3, OH radical, and N radical which generated by the silent electric discharge in the inside of exhaust gas, and the process of oxidation reaction of PM (Carbon subject) is collectively shown in Table 1.

2.4. Parameters which influence PM removal performance seen from the equivalent circuit of Silent Discharge

If the structure and the electric circuit of an electric discharge basic cell of Fig. 8 are transposed to an equivalent circuit here, it will become as it is shown in Fig. 9.

From an equivalent circuit, the relation between applied voltage (Vd), the voltage of a dielectric (Vc), and the voltage between electrode gaps (Vg) is as follows.

$$Vd = Vc + Vg \tag{1}$$

$$V_c = \frac{I}{\omega Cc} \tag{2}$$

$$V_g = \frac{I}{\omega Cg} \tag{3}$$

I is the current which flows into a circuit, and omega(ω) is angular velocity ($\omega = 2\pi$ f), f is the power supply frequency (kHz) and almost constant at around 5 kHz this time. Moreover, the electric capacity of a dielectric (Cc) and an electrode gap (Cg), as follows, respectively.

$$Cc = \frac{\varepsilon \, \varepsilon_0 S}{t} \tag{4}$$

$$Cg = \frac{\varepsilon_0 S}{d} \tag{5}$$

It is here,

S; area (m^2) of a dielectric (Sc) and an electrode (Sg) (S and Sc are nearly equal to Sg)
t; thickness of a dielectric(t) , this time fixed at t= 1 mm
d; electric discharge space gap length
ε_0 ; dielectric constant in a vacuum (ε_0= 8.854x10^{-12} F/m)
ε; relative dielectric constant (the ceramic board used in this research isε= 8.5)

Next, the following relation is realized from between a formula (2) to a formula (5)

$$\frac{Vc}{t} = \frac{I}{\omega \varepsilon \varepsilon_0 S} \tag{6}$$

$$\frac{Vg}{d} = \frac{I}{\omega \varepsilon_0 S} \tag{7}$$

A formula (6) and a formula (7) are showing the electric field strength in a dielectric and electric discharge space. In order to remove PM efficiently through the silent electric discharge which is the purpose of this research, it is necessary to raise the electric field strength (Vg/d) in electric discharge space. Therefore, in order to enlarge Vg from a formula (1) first, it is direct to enlarge the applied voltage Vd itself. However, since it is considered as Vd regularity this time, it is required to make Vc small. From a formula (2) and a formula (4), if *I* and ω (i.e. f) are fixed case , in order to make Cc larger, it is more effective to enlarge relative dielectric constantε. Although the ceramic board (AL2 O3) ε= 8.5 is used this time, if the titanium dioxide (TiO2) ε= 86 and barium titanate (BaTiO₃) ε= 2900, etc. are used, it turns out that improvement in still larger electric field strength is expectable. However, this time research was advanced using the cheap ceramic board which is generally marketed. Moreover, from a formula (7), when Vg is fixed voltage, it turns out that it is effective to make d small and to make electric discharge electrode area S small. From these results of analyzing parameters, the evaluation of discharging characteristics of the electric discharge cell and SDeDPF as opposed to change of the applied voltage Vd were carried out. Then it was considered that discharge gap d and the area of discharging electrode S which influence the performance of electric discharge greatly were important parameters for the evaluation.

3. Experimental devices and method

3.1. Experiment evaluation devices of discharging characteristic with basic structure of SDeDPF

The evaluation equipment of the basic cell of SDeDPF is shown in Fig. 10. Electric supply equipment (High Voltage Pulse Power Supply) is PPS-4000S type custom-made item made from ECG-KOKUSAI. The voltage set up with electric supply equipment from former power supply AC100V is amplified as primary voltage to the maximum voltage of 2 kV, the maximum supply current of 150 mA, and the maximum frequency of 5 kHz, and electric discharge is generated by eventually applying the high frequency and the high voltage of a maximum of 25 kVpp (secondary voltage; i.e. voltage between B and F) to the test sample. As a formula (7), since electric field strength is in discharge current and proportionality relation, in order to measure the current (Ig) between D and E as Fig. 10, 1.5Ω shunt resistance is arranged and current conversion from the amplitude measurement in an oscilloscope is carried out . The size of the discharge current (Ig) is made into the alternative characteristic of electric field strength. A measurement condition is in room temperature and a static state and the preset value of electric supply equipment is the maximum voltage of 0.5 kV and the frequency of 4.8 kHz. When Ver. 6 of the final specification was in an electric discharge state, reduction level of the current value (Ig) was seen about 5% by the result of having passed 2 m³/h air flow rate as a dynamic state at room temperature. It was judged that there was no big influence in evaluating the characteristic of basic structure by static condition with the equipment like Fig. 10.

3.2. Characteristic of high voltage pulse power supply

The result that have been investigated the characteristic of the output voltage of the High Voltage Pulse Power Supply (HVPPS) is shown in Fig.11 and 12. The oscilloscope was directly connected to the last output end (B point and F point in Fig.10) of the HVPPS and the characteristic of voltage and current of pulse power supply itself were measured. Power supply frequency was fixed to f = 5.0 kHz which is the maximum of specification of this device and the change of power supply(1st order = Ep) current and the relation of output (2nd order) voltage are shown in Fig.11. But 150 mA is the maximum supply current on specification, the supplied current serves as a limit by a security apparatus at 130 mA.

Therefore maximum output voltage is 13kVpp at 0.35kv voltage and 130 mA of primary current arranged by Pulse Power Supply. The result in case of changing power supply frequency with fixing the primary voltage 0.5 kV is shown in Fig.12. In this device, the limit that an electric discharge state is stably maintainable was set to about 4.8 kHz and then he output (2nd order) voltage at this time was 24 kVpp without depending on frequency. The voltage waveform (from 23.2 to 24kVpp) outputted from the Pulse Power Supply when it sets to the primary voltage of 0.5 kV and frequency 4.8 kHz is shown in Fig.13 and 14.

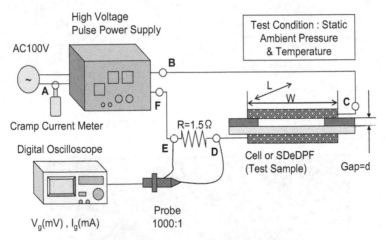

Figure 10. Test equipment for discharging characteristic of SDeDPF or Cell

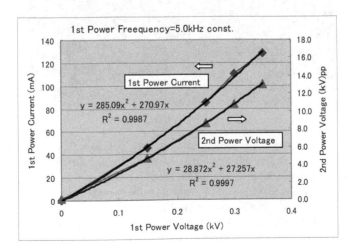

Figure 11. Characteristic of Power Supply (f=5.0kHz)

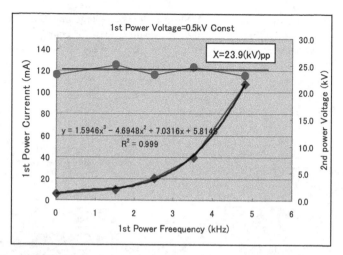

Figure 12. Characteristic of Power Supply (E_p=0.5kV)

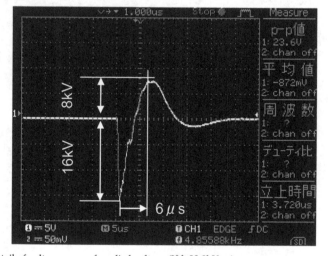

Figure 13. Detail of voltage wave of applied voltage (Vd=23.9kVpp)

Although this Pulse Power Supply (PPS) can switch the polarity (plus or minus) of a pulse arbitrarily, in this research, minus pulse output has been applied. The detail of voltage wave of the impressed voltage (out put voltage from PPS) is shown in Fig. 13, it have reached in 6μs from (-) peak 16kV to (+) peak 8kV. And then the Voltage on SDeDPF is 12kVpp as shown in Fig.14 which is a similar wave to Fig.13. If the above result is summarized, it will become as it is shown in Table 2. With the setting voltage (Ep) and frequency (f) in the PPS, the voltage of Vd (kVpp) will be impressed to SDeDPF. Voltage impressed to SDeDPF is indicated to expressing by Vd(kVpp) after this.

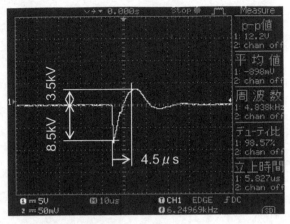

Figure 14. Detail of voltage wave on SDeDPF (Vd=23.9kVpp)

	Setting voltage: Ep (kV)	0.00	0.15	0.25	0.35	0.50
Setting voltage with Pulse Power Supply	Setting Feq.: f (kHz)	0.0	5.0	5.0	5.0	4.8
Applied Voltage to SDeDPF Vd (kV)pp		0.0	4.7	8.7	13.1	23.9

Table 2. Relationship between setting voltage by power supplying device and added voltage on SDeDP

4. Test result

4.1. Test result of discharging characteristic with basic cell and SDeDPF structure

4.1.1. Influence of gap length

The influence of gap length d exerted on electric discharge with the basic cell structure of Fig. 1 in the equipment of Fig. 10 was shown in Fig. 15. The discharge current Ig showed the maximum between d = 1mm to 2 mm. Moreover, looking hard at the state of electric discharge, at not less than d = 2 mm, all the dotted micro discharge becomes strong a spark discharge. In d = 3 mm, the stronger spark discharge in the shortest distance part between electrodes break out and in d = 5 mm , the creeping discharge which the discharge current flows on the spacer wall surface which serves as the shortest distance further are seen and also the noise of the spark discharge became larger. The state of the electric discharge state in between d = 1mm to 2 mm is very stable. In this research of first stage, the basic gap length of SDeDPF was set d =1mm constant which had the maximum discharge current Ig with impressed voltage of 13.1kVpp, and that was considered the power saving by 13.1kVpp rather than 23.9kVpp.

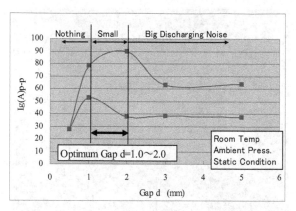

Figure 15. Influence of Gap

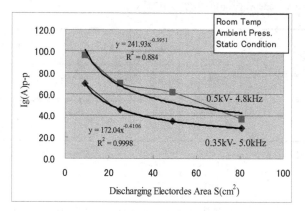

Figure 16. Influence of Electrode Area

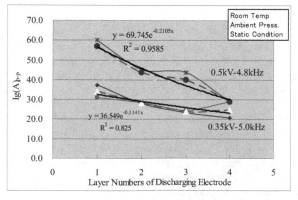

Figure 17. Influence of Layered Numbers of Electrodes (Total Area)

Type	SDeDPF Cell Layers	Discharge Gap d (mm)	Electrode Position D (mm)	Electrode Length L (mm)	Total Electrode Area S (cm²)	Electrode Position Numbers	Turbulent Blocks
Ver.2	2	1.0	22	70	98	1	No
Ver.3	4	1.0	47	20	56	1	No
Ver.5	8	1.0	52	10	56	1	No
Ver.6	8	1.0	52	10	56 (50.6)	1	With

Table 3. Each Type of SDeDPF for Evaluation of PM Removal

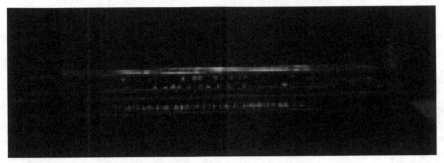

(Violet-bluish Micro-streamer Discharge, white part is lighting)

Figure 18. Example of Discharging of SDeDPF (Ver.3 Vd=13.1kVpp)

4.1.2. Influence of discharging electrode area

With the equipment of Fig.10, the result of having checked the relation of electrode area (S) and the discharge current (Ig) is shown in Fig. 16. Four variations of discharging electrode area (S) were prepared, based on 70x70 mm, 90x90 mm, 50x50 mm, and 30x30 mm.

Discharge current (Ig) is converted from voltage measurement of the oscilloscope in the 1.5 Ω shunt resistance between D and E of Fig. 10.

As shown in Fig. 16, if the discharging electrode area (S) becomes large, the discharge current (Ig) will decrease. It is the reason that as understanding from a formula (7), if the discharging electrode area (S) increase, then the electric field strength will fall, and sufficient discharge current (Ig) is not acquired. In order to secure an exhaust pressure loss equivalent to the present DPF made from ceramics, it turns out that it is necessary to reduce the total area of electrodes. Because the electric field strength falls, if it does not change the frontage size (W) of SDeDPF entrance, it is necessary to increase the number of stages of the electric discharge basic cell which will be laminated.

4.1.3. Influence of electrode layer numbers

As mentioned above, it is expected that the one which has larger discharge current (Ig) has the higher PM removal performance. If the lamination number of sheets of an electrode becomes increase (the total electrode area increase), electric field strength will fall from a formula (7) and then PM removal performance declines. Fig. 17 shows the relation of the number of lamination stages and the discharge current (Ig). It is the result of laminating W = 70 mm and L = 70-mm MFS electrode from 1 to 4 layers. If the discharge current (Ig) will be four layers, it will be reduced by half compared with one layer. If the frontage of W = 70 mm is fixed, and for making discharge current equivalent, it necessary to set L to one half. Furthermore, an exhaust pressure loss is considered, for considering it as eight layers, it is necessary to set L to one fourth. Ver. 2 to Ver. 6 is manufactured as specification which evaluates PM removal performance to the above electric discharge characteristic as shown in the examination result of an influence factor to the table 2. Ver. 6 was made into the last specification of this stage.

Fig. 18 is an example of the state of discharging of SDeDPF type of Ver.3 which Vd=13.1kVpp was impressed. Violet-bluish many micro-streamer discharge were seen.

4.1.4. Test Equipment of PM removal mechanism analysis

Until now, the parameter which influences PM removal was clarified based on PM removal mechanism by the silent electric discharge described by Section 2.3 and 2.4 and improvement in PM removal ratio has been aimed at by various kinds of variation trial productions. As a result, PM removal ratio which was almost quite high as prediction and examination has been attained.

However, it needed to verify whether really such a PM removal mechanism would act, and the verification was performed with experimental devices as shown in Fig.19. PM (carbon subject) which supplied from PM generating equipment (PM Generator) was oxidized from C to CO_2 or CO by O, O_3, OH and N radical which were strong activated oxygen sorts and generated by the plasma which occurred in SDeDPF. And then in the exit of SDeDPF, it is measured how CO or CO_2 concentration changes.

PM generating equipment (PM Generator) is mentioned in detail at next section 4.3 and high frequency and high voltage supply unit (Pulse Power Supply) are completely the same as what was explained in Fig.10. The exclusive probe of CO and CO_2 concentration measuring device (IAQ Monitor Model2211 made from CO&CO_2 Meter; KANOMAX) is inserted in the exit of SDeDPF, and as opposed to change of the applied voltage to SDeDPF , the change of CO and CO_2 concentration is measured.

4.2. Verification result of PM removal mechanism

As a verification of PM removal mechanism, the measurement result of concentration change of CO and CO_2 as opposed to the applied voltage to SDeDPF is shown in Fig. 20. Although CO&CO_2 of the entrance of SDeDPF is the concentration of the usual ambient level, it seems that a little since an exit is in the state where PM is supplied from PM generating equipment, CO_2 by that of combustion are going up. As the base of this point, the concentration change of CO_2 and CO was measured when the applied voltage to SDeDPF is raised.

It has confirmed CO_2 concentration rising, and C oxidizing to CO_2 as opposed to increasing the applied voltage Vd. In particular, in maximum applied voltage Vd=23.9kVpp of this equipment, the rapid increase in CO_2 and also CO can obtain the data of a rapid upward tendency, and can guess that the oxidation reaction from C to CO_2 or CO is advancing violently.

From this result, it was judged that between the applied voltage Vd on SDeDPF and concentration change of CO_2 and CO was interlocking closely, and that the mechanism by which PM was oxidized and removed with a strong activated oxygen sorts by silent electric discharge was supported.

In the applied voltage Vd from 0 to 23.9kVpp, the average increase of CO_2 concentration is about 10 ppm (from 390 ppm to 400 ppm), and similarly although the average increases of CO concentration are few, it is about 1 ppm (from 0 to 1 ppm).

Furthermore, in order to support the above-mentioned change of state, the increase in mass as opposed to increase of CO_2 and CO concentration was calculated by using the gas equation of an equation (8). The content of Carbon (C) which was a main ingredients of PM was further calculated from those molecular weight ratios.

$$V = (w/M)RT/P \tag{8}$$

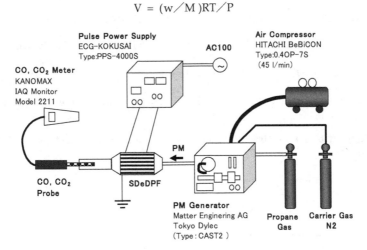

Figure 19. Test equipment of Measuring CO_2 &CO conversion from C (SDeDPF; Ver.6)

Here,

P: pressure (atom) =1, V: volume (liter), w: mass of CO_2 or CO,
M: the molecular weight of CO (28g) or CO_2 (44g), R: constant gas factor =0.0821,
T: absolute temperature (K) = 25+273.15 = 298.15 (constant as a room temperature)

Since 1m³ is 1000 liters and the volume ratio of 1/1000000 is set to 1 ppm, the volume of 10 ppm CO_2 is set to $V_{CO2} = 10 \times 10^{-3}$ (liter) from following calculation.

$$\left(V_{CO2}\ /\ 1000\right)\ /\ \left(1\ /\ 1000000\right)\ =\ 10ppm \quad i.e.\ V_{CO2} = 10 \times 10^{-3}\left(liter\right) \tag{9}$$

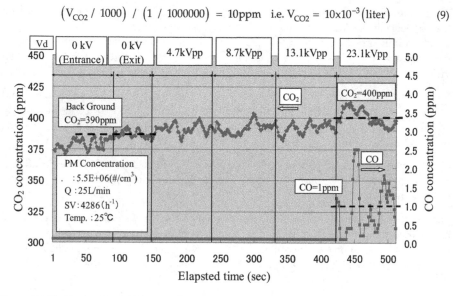

Figure 20. Verification Test of Changing from C to CO and CO_2 by Silent Discharge

Similarly, the volume of 1 ppm CO is set to $V_{CO} = 1 \times 10^{-3}$ (liter) from following calculation.

$$\left(V_{CO}\ /\ 1000\right)\ /\ \left(1\ /\ 1000000\right)\ =\ 1ppm \quad i.e.\ V_{CO} = 1 \times 10-3\left(liter\right) \tag{10}$$

When these results of calculation (10) and (11) are calculated by a formula (8), increase mass of CO_2: w_{CO2} is following.

10×10^{-3} (liter) = (w_{CO2} / 44) × 0.0821 × 298.15 / 1.0   i.e. w_{CO2} = $(10 \times 10-3 \times 44)$ / 0.0821×298.15 = 17.98×10^{-3} (g)

Therefore, the content of C is from a molecular weight ratio,

$$17.98 \times 10^{-3}x\ (\ 12 / 44)\ =4.9 \times 10^{-3}\left(g\right) \tag{11}$$

Similarly, an increase mass of CO: w_{CO} is following.

1.0×10^{-3} (liter) = (w_{CO} / 28) × 0.0821 × 298.15 / 1.0   i. e. w_{CO} = $(1.0 \times 10-3 \times 22)$ / 0.0821×298.15 = 1.15×10^{-3} (g)

Therefore, the content of C is from a molecular weight ratio,

$$1.15 \times 10^{-3}x\ (\ 12 / 28)\ =0.49x\ 10^{-3}\left(g\right) \tag{12}$$

The increase sum total of C is set to 5.39×10^{-3} (g) from the above calculation result (11) and (12). Since the premise is calculation by of 1 m³, this is replaced with 5.39×10^{-3} (g/m³).

On the other hand, the result of the amount of mass change directly measured with PM number concentration analyzer at the time of silent electric discharge with Ver.6 is 5.48×10^{-3} (g/m³). Therefore, it has confirmed certain similarity between the calculated value by the gas equation of state (8) and the direct measured value with PM number concentration analyzer. From this result, it was proved quantitatively that the concentration change of CO_2 and CO at the time of the electric discharge respond to PM oxidizing and increasing, and moreover I think certainly that PM removal mechanism by silent electric discharge was supported quantitatively.

4.3. Test Equipment for PM Supply System (PPM) and PM Removal Evaluation

Test Equipment of PPS and PM Removal Evaluation is shown in Fig. 21. A fixed concentration PM which was generated by PM Generator (combustion particulate generating equipment) was supplied to SDeDPF. Raw Gas Dilution & Engine Exhaust Particle Sizer (an exhaust gas particulate number counting system) measures the change of PM concentration as opposed to the change of the high frequency and the high voltage which was impressed to SDeDPF and evaluates the PM removal rate form the decreasing change of the PM concentration. PM is generated when adjusted N_2, the air content and the propane gas were mixed and burned in PM Generator. Here, PM concentration means the particulate number (#/cm³) which contained per sampling gas 1cm³. The amount of supplying gas is 0.25 L/min (1.5m³/h) and SV of SDeDPF at this time is a maximum of 4286 (h⁻¹).

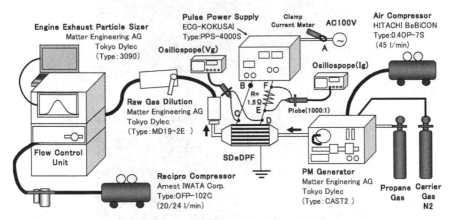

Figure 21. Test equipment for PM Supply System and PM Removal Evaluation

Fig. 22, 23, 24 and 25 are coincidence measurement results of the voltage (Vg) (between C and D in Fig. 21) and discharge current (Ig) (between E and F in Fig. 21) which occurs in SDeDPF when Vd =23.9kVpp is impressed by a Pulse Power Supply in the measurement circuit of Fig. 21. As shown in Fig. 22, the charge-and-discharge current of the corresponding phase in sync with the frequency of 4.8 kHz of service voltage has occurred in pulse. Fig. 23 and 24 are the waveform which expanded Fig. 22. Each two of Non-discharging period and Discharging period [16][17] exist in one cycle, respectively.

Furthermore, the waveform which expanded Fig. 24 is shown in Fig. 25. At first discharge, the minus pulse voltage Vg rises and after 0.5 microsecond of non-discharging periods, charge-and-discharge current (a max. of 157 App) generates in the pulse of high frequency at the starting discharge Voltage Vs = -2.5kV and then the discharging period continues for 0.8 microsecond. Moreover the 2nd discharging period has passed the non-discharging period for 3.6 microseconds since the 1st discharging period as shown in Fig. 24, in charge-and-discharge current (a max. of 73 App), the pulse of high frequency continues for 4 microseconds.

Then the current measured with clamp meter was 0.88A in the AC100V line in front of the PPS unit at this time, so it was confirmed that the electric power was 88W including the PPS and SDeDPF.

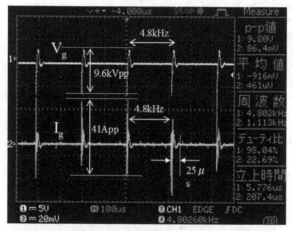

Figure 22. Charge & Discharge Voltage (Vg) and current(Ig) (Vd =23.9kVpp)

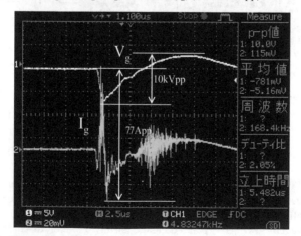

Figure 23. Detail of Figure 21 (1 div. =2.5 μ s)

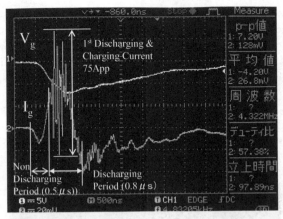

Figure 24. More detail of Figure 22. (1 div. =1.0 μ s)

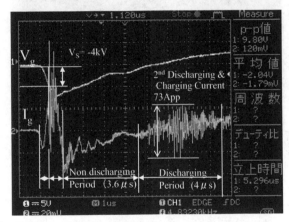

Figure 25. More detail of Figure 23. (1div. =500ns)

4.4. Electric power consumption

The average electric power consumption of the whole equipment including PPS unit and SDeDPF was calculated from the clamp current measurement value in the A section of Fig. 12. It has measured in the state of voltage Vd=23.9kVpp currently applied to SDeDPF. As an AC100V-50Hz was original power supply to PPS, the voltage of Ep=0.5 kV, 4.8 kHz. was setting in the PPS unit of this research.

As a result of measuring using final Ver. 6, the effective value voltage in Vrms =100(V) and clamp current measurement of the effective value current in the A section was Irms=0.88 (A), and it turned out that effective (apparent) electric power (Pr) was 88(W).

On the other hand, as shown in Fig. 26, electric power consumption was calculated by the base of the coincidence measurement result of the voltage (Vg) and the current (Ig) of

SdeDPF Ver. 6, when Vd =23.9kVpp was applied on it. Momentary power consumption (Wt) was calculated from multiplying current value (It) simultaneous by the voltage value (Vt) measured for every microsecond, the integrating amount of electric power consumption was divided by a cycle T = 208 μ s and then average electric power consumption (Pr) was calculated like a formula (10) when expressed to the formula, and obtained the result of average electric power consumption Pr =90 (W) as shown in Fig. 27. In formula (13), they are f = 4.8 kHz, T= 208 μ s, and dt =1 μ s. Therefore, it becomes power consumption almost equivalent to the clamp current measurement result of 88 (W), and it can be said that the power consumption of the SDeDPF Ver.6 of this research is before and after 90 (W).

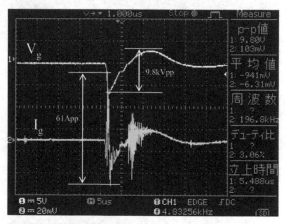

(Voltage (Vg), Current (Ig) at Vd =23.9kVpp with Ver.6)

Figure 26. Base data for calculation of power consumption

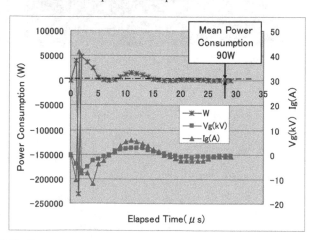

(Vd =23.9kVpp with Ver.6)

Figure 27. Calculation result of power consumption

$$\Pr = \int_0^T \left(V_t \cdot I_t \cdot dt \right) / T \qquad (13)$$

4.5. Evaluation result of PM removal

Fig. 28 and 29 are PM removal quality assessment result under the room temperature and atmospheric pressure conditions of last specification Ver.6. SDeDPF was equipped as shown in Fig. 21, progress of PM concentration (#/cm³) change as opposed to changing the applied voltage by PPS unit was shown in Fig. 28. The relationship of PM concentration and distribution of the diameter of PM particle (Particle Size (nm)) in the stable point on each condition [1] to [6] of Fig. 28 are shown in Fig. 29. Supplied PM concentration ([1] PM Direct) from PM generating equipment is made into a starting point. It is the result of changing applied voltage and frequency in order of [3] to [6] after connecting SDeDPF to PM generating equipment ([2] Put ON; power supply OFF). At the condition [6] Vd =23.9kVpp, PM removal ratio of 95.6% is attained as opposed to the condition of the [2] power supply OFF.

Moreover, as shown in Fig. 29, sufficient reappearance has been checked from the repetition result of [6] power supply ON - [7] OFF - [6] ON - [7]OFF. Although in an early stage, 13.8% of PM adhesion in SDeDPF from the difference of the conditions [1] and [2] is checked, also after the condition [3], on while the voltage was impressed, it turned out that PM was removed continuously and efficiently even under room temperature and atmospheric pressure conditions. Although in Fig. 28, PM grain size distribution does not change a distribution state so much focusing on 100 nm of it, the absolute quantity of PM concentration is decreasing sharply.

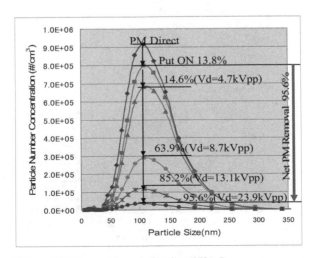

Figure 28. Elapsed Time of PM Removal by each Condition (Ver.6)

In Fig. 30, it is the result of carrying out comparison arrangement of the PM removal performance of each specification to applied voltage Vd. Especially, Ver.6 with a turbulent flow block (T. B) have improved PM removal ratio of not less than 15% as opposed to nothing Ver5 mostly in all the voltage regions. The effect of T.B was checked certainly.

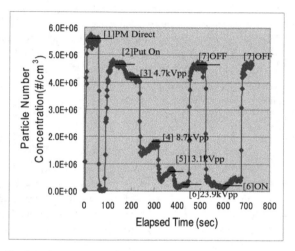

Figure 29. Particle Size Distribution at each Condition (ver.6)

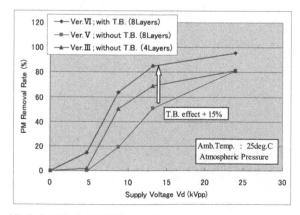

Figure 30. Effect of Turbulent Block for PM Removal

4.6. Relationship between PM removal quantity and pressure loss

The horizontal axis of Fig. 31 is a pressure loss of SDeDPF which is measured by blower fan equipment. It is checked by 4 m ^{3}/min of air mass flow. Although the pressure loss of SDeDPF Ver.3 (4 Layers) which stacked four steps of basic cells was 5kPa, the pressure loss of SdeDPF Ver.5 (8 Layers) which stacked eight steps of basic cells was about 1kPa. As a result, it was set to one fifth and set to the level equivalent to the current ceramics DPF. And although SDeDPF Ver.6 added the Turbulent Block to Ver.5, since it was not the structure which closed a channel completely, it secured pressure loss almost equivalent to Ver.5. Ver.3 (4Layers) of pressure loss was decreased by 8Layers which had twice about the passage cross-sectional area of exhaust gas and PM removal ratio was maintained by making equivalent electric field strength (total electric discharge area) with shortening electrode length from L= 20 mm to 1 mm further. Furthermore, Ver.6 was added the Turbulent Block (T.B.) to the electric discharge space part of Ver.5, but the pressure loss was able to keep as almost equally and it was able to raise PM removal ratio from 81.35% to 95.6%.

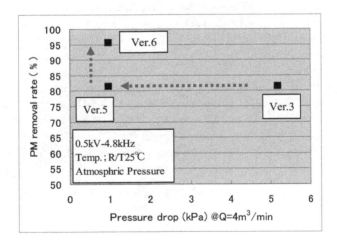

Figure 31. Relationship of PM Removal and Pressure Drop of SDeDPF

4.7. Relationship between PM removal quantity and power consumption

In Fig. 10, from the current measurement result of AC100V line with the clamp meter, when the voltage 23.9kVpp was applied on SDeDPF Ver.6, the electric power consumption of Ver. 6 was 88W. Moreover, from the current measurement result of having inserted 1.5Ω resistance in this line, it was 84W as almost equally as the result of clamp meter. On the other hand, In PM removal evaluation of Fig. 28, the initial average number of PM concentration injected to SDeDPF Ver.6 was 5.5E+06 (#/cm^3) and then it had been reduced to

2.1E+05 (#/cm^3) by passing through SDeDPF Ver.6 with electric discharge impressed 23.9kVpp voltage. When these average number of PM concentration were converted into the PM weight concentration, it is equivalent to the quantity from 5616(μ g/m^3) to 185 (μ g/m^3). Then the supplied gas flow rate was 25 L/min (1.5m^3/h) which was containing PM from PM Generator. Therefore, the quantity: W of PM removed by electric discharge per hour (g/h) is set to following.

$$W = (5616 - 185) \times 1.5 = 8.15 \times 10^{-3} (g / h) \tag{14}$$

Since power consumption is 84w, the amount of PM removal (processing):W_{1kwh} per 1kwh was set to following.

$$W_{1kwh} = 8.15 \times 10^{-3} / 84 \times 10^{-3} = 0.09 (g / kWh) \tag{15}$$

5. Conclusion

1. Silent Discharge type of electric DPF (SDeDPF) of this research is characterized by applying the unique MFS electrode and having Turbulent Brock in the lamination type of eight layers which has almost same exhaust pressure loss as current DPF made by ceramics. Finally, PM removal ratio of 95.6% has been attained with power saving of 84 to 90W also under using no precious-metals, and room temperature and atmospheric pressure conditions.
2. Although the amount of PM removal (processing) per 1kwh became 0.09 g/kWh, the dielectrics used for SDeDPF of this research was a ceramic board which the dielectrics constant was $\varepsilon = 8.5$, generally marketed in, cheap, and easy to come to hand. If the titanium dioxide (TiO$_2$) $\varepsilon = 86$ or barium titanate (BaTiO$_3$) $\varepsilon = 2900$ etc. were used, it was planed improvement in large electric field strength, and then the amount of PM removal might be further improvable.

I think that PM removal potential with SDeDPF in the minimum specification of a dielectric was able to be clarified this time.

Author details

Minoru Chuubachi and Takeshi Nagasawa
Utsunomiya University, Japan

6. References

[1] Kanesaka H., Yoshiki H., Tanaka T., Akiba K., (2001) "Some Proposals to Low-Emission, High-specific-Power Diesel Engine Equipped with CRT", SAE2001011256, pp.1-8

[2]   CRT,   CCRT,   "Large-sized   Diesel   Catalyst   Division",   Johnson   Matthey,
      http://www.jmj.co.jp/diesel/crt.html , Accessed 2012 Mar. 28
[3]   Daisho Y., (2006) "Recent Trends on Research and Development for Improving Motor
      Vehicle Exhaust Emission and Fuel Economy", Denso Technical Review, Vol.11, No.1,
      pp3-9
[4]   Hori M., (2006) "Future Prospect of Eco-Diesel Engine", JSAE of Japan, Vol.60, No.9
      (2006), pp.6-11
[5]   Hirata K., Masaki N., Akagawa H., (2006) "The Urea-SCR System for Heavy-Duty
      Commercial Vehicle", JSAE of Japan, Vol.60, No.9, pp.28-33
[6]   Chuubachi M., Nagasawa T., (2009) "Feasibility Study of a Silent Discharge Type of
      DPF Applied the Metal Fiber Sheet for an Electrode", Paper Book of JSAE Kantou-
      Block Joint Lecture Meeting in 2009 Maebashi, OS0702, pp.3-4
[7]   Chuubachi M., Nagasawa T., (2010) "Feasibility Study of a Silent Discharge Type of
      DPF without Precious Metal under Room Temperature and Atmospheric Pressure
      Condition", Transactions of the Japan Society of Mechanical Engineers B , Vol.76,
      No.772, pp.292-294
[8]   Yukimura K., (2008) "Discharge Plasma Engineering", EEText, First edition, IEEJ, pp.32-
      33
[9]   Yao S., (2009) "Plasma Reactors for Diesel Particulate Matter Removal", Recent Patents
      on Chemical Engineering, No.2, pp.67-75
[10] Kim Y.H., et. Al, (2009) "Non-Thermal Plasma PM Removal System for Diesel
      Passenger Vehicles", JSAE Annual Congress (Fall), Proceedings, No.89-09, No.12-
      20095727
[11] Kogelschatz U., (1999) "From Ozone Generators to Flat Television Screens: History and
      Future Potential of Dielectric-Barrier Discharges", Pure Apple. Chem., Vol. 71, No.10,
      pp.1819 – 1828
[12] Harano A., (1998) "Oxidation of Carbonaceous Particles in Silent Discharge Reactor",
      Journal of Chemical Engineering of Japan, Vol.31, No.5, pp.700-705
[13] Yao S., (2007) "Comprephensive Technological Development of Innovative,
      Next-Generation, Low-Pollution Vehicles, Basic Study of PM Oxidation Promoted
      by O_3 and NO_2 ", JSAE Annual Congress (Fall), Proceedings, No.87-07, No.12-
      20075799
[14] Tanaka S., Takakura S., Matsubara S., et. Al, (2006) "A Study on Low Temperature
      Oxidation of PM in Diesel Exhaustby using Non-thermal Plasma", Transaction of JSAE,
      Vol.37, No.6, pp.73-78
[15] Yukimura K., (2008) "Discharge Plasma Engineering", EEText, First edition, IEEJ,
      pp.132-137
[16] Tamita T., Iwata A., Tanaka M., "Discharge Measurement of ac Plasma Display panels
      using V-Q Lissajous Figure", IEEJ Transactions A, Vol.118, No.4, pp.353-358

[17] Tanaka M., Yagi S., Tabata N., (1982) "Observation of Silent Discharge in Air" IEEJ Transactions A, Vol.102, No.10, pp. 9-16

Analytical Methodologies for the Control of Particle-Phase Polycyclic Aromatic Compounds from Diesel Engine Exhaust

F. Portet-Koltalo and N. Machour

Additional information is available at the end of the chapter

1. Introduction

Diesel emissions contain complex mixtures of chemical constituents that are known to be (or possibly to be) human carcinogens, or that have adverse health effects [1]. Among these substances (formaldehyde, benzene, acrolein, dioxins, etc.), polycyclic aromatic hydrocarbons (PAHs) and their nitrated derivatives (nitro-PAHs) have been implicated as major contributors to the toxicity of diesel exhausts [2, 3]. However, these substances are currently non-regulated pollutants in engine exhaust. PAHs are a group of organic compounds consisting of two or more fused aromatic rings, which are produced as a result of the incomplete combustion of fossil fuels. These compounds are mainly derived from anthropogenic sources, particularly from mobile sources of emissions in urban areas [4]. The combustion of fuel in diesel engines results in the formation of a mixture of gaseous compounds (CO, CO_2, NO, NO_2, SO_2, etc.) and solid particles (carbonaceous matter, sulphates, trace metals, etc.). PAHs and their derivatives can be found in the gaseous or particulate phases of diesel exhaust fumes depending on their vapour pressure, but the distribution of particles between the two phases is also dependent on the total amount of particles, on the physical characteristics of the particles (particularly their specific surface area), and on the temperature.

The majority of particulate matter in diesel exhaust is composed of fine particles that are primarily formed from the condensation of organic matter on an elemental carbon core. These particles are generally called soot particles. The soluble organic fraction (SOF) is defined as the fraction that can be extracted from soot using an organic solvent; particulate PAHs and their derivatives are found in this fraction. The SOF primarily originates from unburned fuel and engine lubrication oil [5]. There is a wide range of SOF in the diesel particulate matter; it can range from less than 10 % to more than 90 % by soot mass. The fraction of SOF in the particulate matter depends on the type of diesel vehicles (light or heavy-duty diesel vehicles, mopeds, etc.), on the concentrations of sulfur and aromatic compounds in the diesel fuel, on the test engine cycles and also, on the collecting procedure of diesel exhaust, which is less well described. In general, the SOF values are highest for light engine loads when exhaust temperatures are low [6].

Two methods are generally used for measuring vehicle emissions: dynamometer studies where tailpipe fumes are measured from one vehicle, and roadside or tunnel studies where mean traffic emissions values (daily, by vehicles, etc.) are obtained [7]. In this book chapter, only analytical methodologies developed for diesel emissions studies using dynamometer tests are described. Even if the exhaust particles produced in the tailpipe are distinct from those sampled in the ambient atmosphere, it is shown in this chapter that the trapping media are very similar. Conventional and less conventional sampling trains for collecting the particles are described. The classical (Soxhlet and ultra-sound extractions) and more recent methodologies (micro-wave assisted extractions, pressurised solvent extractions, supercritical fluid extractions, and solid phase extractions) are detailed; they are used for treating the collected samples from gaseous and particulate phases to obtain the fraction containing PAHs and their derivatives. Because the individual products present different health risks, information about total PAH emissions is less important than information about the composition of such emissions. Therefore, techniques that are developed for identifying and quantifying the individual PAHs and their nitrated derivatives found in the two phases are described.

Finally, the efficiency of the sample pretreatment step (before quantitation) is strongly dependent on the SOF fraction of diesel soot; therefore, it is shown in this chapter that when the optimisation of this important analytical step is neglected, it can lead to important analytical bias.

2. Properties, sources and toxicity of PAHs and nitro-PAHs

2.1. Physical and chemical properties of PAHs and nitro-PAHs

The name "polycyclic aromatic hydrocarbons" (PAH) commonly refers to a large class of organic compounds containing two or more fused aromatic rings; however, in a broad sense, non-fused ring systems should also be included. In particular, the term "PAH" refers to compounds containing only carbon and hydrogen atoms (*i.e.*, unsubstituted parent PAH and their alkyl-substituted derivatives), whereas the more general term "polycyclic aromatic compounds" also includes the functional derivatives (*e.g.*, nitro- and hydroxy-PAHs) and the heterocyclic analogues, which contain one or more hetero atoms in the aromatic structure (aza-, oxa-, and thia-arenes).

The physical and chemical properties of PAHs are dependent on the number of aromatic rings and on the molecular mass (Table 1). The smallest member of the PAH family is naphthalene, a two-ring compound, which is mainly found in the vapour phase in the atmosphere, because of its higher vapour pressure and Henry's law constant K_H (ratio between the partial pressure of a gas above a solution and the amount of gas solubilized in solution). Three to five ring-PAHs can be found in both the vapour and particulate phases in air. PAHs consisting of five or more rings tend to be solids adsorbed onto other particulate matter in the atmosphere. For instance, the resistance of PAHs to oxidation, reduction, and vaporisation increases with increasing molecular weight, whereas the aqueous solubility of these compounds decreases (their n-octanol/water partition coefficient log K_{ow} is higher). As a result, PAHs differ in their behaviour, their distribution in the environment, and their effects on biological systems.

Compounds/ Abbreviations	Chemical structure	Molecular mass (g mol^{-1})	Melting/ Boiling points (°C)	Vapour Pressure (Pa at 25°C)	Log K_{ow}	Solubility in water at 25°C (mg L^{-1})	K_H at 25°C (Pa.m^3 mol^{-1})
Naphthalene* NAPH		128.18	81 / 218	10.4	3.4	31.7	49
Acenaphtylene* ACY		152.18	92 / 270	0.89	4.07	-	114
Acenaphthene* ACE		154.21	95 / 279	0.29	3.92	3.9	15
Fluorene* FLUO		166.22	116 / 295	0.08	4.18	1.68	9.81
Anthracene* ANT		178.24	216.4 / 342	8.0 10^{-4}	4.5	0.073	73
Phenanthene* PHEN		178.24	100.5/ 340	0.016	4.52	1.29	4.29
Fluoranthene* FLT		202.26	108.8/ 375	0.00123	5.20	0.26	1.96
Pyrene* PYR		202.26	150.4 / 393	0.0006	5.18	0.135	1.1
Chrysene* CHRYS		228.29	253.8 / 436	-	5.86	0.00179	0.53
Benz[a] Anthracene* B(a)ANT		228.29	160.7 / 400	2.8 10^{-5}	5.61	0.014	1.22
Benzo[b] Fluoranthene* B(b)FLT		252.32	168.3 / 481	-	5.78	0.0015	0.051
Benzo[k] Fluoranthene* B(k)FLT		252.32	215.7 / 480	-	6.11	0.0008	0.044
Benzo[a]pyrene* B(a)PYR		252.32	178.1 / 496	7.3 10^{-7}	6.50	0.004	0.034 (20°C)
Benzo[e]pyrene B(e)PYR		252.32	178.7 / 493	7.4 10^{-7}	6.44	0.005	-

Compounds/ Abbreviations	Chemical structure	Molecular mass (g mol⁻¹)	Melting/ Boiling points (°C)	Vapour Pressure (Pa at 25°C)	Log K_{ow}	Solubility in water at 25°C (mg L⁻¹)	K_H at 25°C (Pa.m³ mol⁻¹)
Perylene PER		252.32	277.5 / 503	-	5.3	0.0004	-
Benzo[g,h,i] Perylene* B(ghi)PER		276.34	278.3 / 545	$1.4\ 10^{-8}$	7.10	0.00026	0.027 (20°C)
Indeno[1,2,3-cd] Pyrene* InPYR		276.34	163.6 / -	-	-	0.00019	0.029 (20°C)
Dibenz[a,h] Anthracene* DB(ah)ANT		278.35	266.6/ -	-	6.75	0.00050	-
Coronene COR		300.36	439 / 525	$2\ 10^{-10}$	5.4	0.00014	-
2-nitronaphthalene 2N-NAPH		173.17	74 / 304	$3.2\ 10^{-2}$	2.78	26	610
2- nitrofluorene 2N-FLUO		211.22	154 / 326	$9.7\ 10^{-5}$	3.37	0.216	95
9-nitroanthracene 9N-ANT		223.23	146 / -	-	4.16	-	-
2-nitrophenanthrene 2N-PHEN		223.23	120 / -	-	4.23	-	-
2-nitrofluoranthene 2N-FLT		247.25	- / 420	$9.9\ 10^{-7}$	-	0.019	13
1-nitropyrene 1N-PYR		247.25	152 / 472	$4.4\ 10^{-6}$	5.29	0.017	64
2-nitropyrene 2N-PYR		247.25	199 / 472	$4.4\ 10^{-6}$	-	0.021	64
7-nitrobenz[a] anthracene 7N-B(a)ANT		273.29	162 / -	-	5.34	-	-
6-nitro benzo[a]pyrene 6N-B(a)PYR		297.31	260 / 567	-	6.13	-	12

* 16 priority PAHs defined by the US-EPA

Table 1. Physical and chemical properties of some PAHs and nitro-PAHs [8, 9].

2.2. Toxicity of PAHs and nitroPAHs

Over 100 chemical compounds formed during the incomplete combustion of organic matter are classified as PAHs. These ubiquitous environmental pollutants are human carcinogens and mutagens; therefore, they are toxic to all living organisms. There are various pathways for human exposure to PAHs. For the general population, the major routes of exposure are from food and inhaled air; however, in smokers, the exposure from smoking and food may be of a similar magnitude. Extensive reviews and guidelines concerning PAHs contamination in food and air have been performed, and they have questioned the correct marker capable of following the risk assessment for the population [10].

There have been concerns regarding the carcinogenic effects of these products for at least two centuries, following the report of a high incidence of scrotal cancer in soot workers in London by Sir P. Pottwatched. The association between cancer and a specific chemical compound, Benz[a]pyrene, was established when this compound was isolated from chimney soot. Several subsequent studies proved that, in addition to Benz[a]pyrene, other PAHs were also carcinogenic [11]. The United States Environmental Protection Agency (U.S. EPA) targeted sixteen specific PAHs for measurements in environmental samples (see Table 1), and Benz[a]pyrene (indicator of PAHs species) was classified as a 2B pollutant, which means that it is a probable human carcinogen based on sufficient evidence from animal studies, but there is inadequate evidence from human studies [12]. According to the International Agency for Research on Cancer Classification (IARC), it is Group 2A compound, which means that it is likely carcinogenic to humans [13]. The World Health Organization (WHO) also added PAHs to the list of priority pollutants in both air and water. France, Japan, Germany, Netherlands, Sweden, and Switzerland established emission standards for most of the hazardous air pollutants, including PAHs. The WHO and the Netherlands even established ambient air quality guidelines for PAHs (1.0 ng m^{-3} and 0.5 ng m^{-3}, respectively). These limits are not legally binding, but the common consensus is that these pollutants require maximum reduction or "zero levels" in emissions.

The increased attention to nitro-PAHs is due to their persistence in the environment and the higher mutagenic (2×10^5 times) and carcinogenic (10 times) properties of certain compounds compared to PAHs [14]. Studies have highlighted the important role of nitro compounds, and they emphasised the necessity of improving primary prevention methods to reduce nitrated molecule air pollution. Today, the known sources of these types of compounds are a variety of combustion processes, especially in diesel engines. It was reported that the main contributors to the direct-acting mutagenicity of diesel exhaust particulates were nitro-PAHs, including 1,3-, 1,6- and 1,8-dinitropyrenes, 1-nitropyrene, 4-nitropyrene, and 6-nitrochrysene [15-18]. Therefore, researchers have investigated the chemical and physical properties of particulate matter in diesel exhaust according to the development of diesel technology and emissions regulation; comparisons between the fuel and engine used, such as diesel versus gasoline, light-duty diesel, biodiesel and any so called "clean diesel" have also been reported [19-23].

2.3. Contribution of mobiles sources to atmospheric emissions.

PAHs primarily originate in heavily urbanised or industrialised regions; therefore, the majority of these compounds are anthropogenic. Different sources of PAHs were reviewed and classified according to five main categories (domestic, mobile, industrial, agricultural and natural) [24]. Domestic emissions are predominantly associated with the burning of coal, oil, gas, garbage, or other organic substances such as tobacco or char broiled meat. Mobile sources include the emission from vehicles such as aircraft, shipping vehicles, railways, automobiles, off-road vehicles, and machinery. Industrial sources of PAHs include primary aluminium production, coke production, creosote and wood preservation, waste

PAH ratio	Value range	Source	References
Σ(low weight PAHs) / Σ(high weight PAHs)	<1	Pyrogenic	[27]
	>1	Petrogenic	
FLT/(FLT+PYR)	<0.5	Petrol emissions	[28]
	>0.5	Diesel emission	
ANT/(ANT+PHEN)	<0.1	Petrogenic	[29]
	>0.1	pyrogenic	
FLT/(FLT+PYR)	<0.4	Petrogenic	[30]
	0.4-0.5	Fossil fuel combustion	
	>0.5	Grass, wood, coal combustion	
B(a)PYR/(B(a)PYR+CHRYS)	0.2-0.35	Coal combustion	[31]
	>0.35	Vehicular emissions	
	<0.2	Petrogenic	[32]
	>0.35	Combustion	
B(a)PYR/(B(a)PYR+B(e)PYR)	~0.5	Fresh particules	[33]
	>0.5	Photolysis (ageing of particles)	
InPYR/(InPYR+B(ghi)PER)	<0.2	Petrogenic	[32]
	0.2-0.5	Petroleum combustion	
	>0.5	Grass, wood, coal combustion	
2-MeNAPH/PHEN	<1	Combustion	[34]
	2-6	Fossil fuels	
Σ MePHE/PHEN	<1	Petrol combustion	[35]
	>1	Diesel combustion	
B(b)FLT/B(k)FLT	2.5-2.9	Aluminium smelter emissions	
B(a)PYR/B(ghi)PER	<0.6	Non traffic emissions	[36]
	>0.6	Traffic emissions	

Table 2. Diagnostic ratios reviewed from [26].

incineration, cement manufacturing, petrochemical and related industries, bitumen and asphalt industries, rubber tire manufacturing, and commercial heat/power production. Agricultural sources include stubble burning, open burning of moorland heather for regeneration purposes, and open burning of brushwood and straw. With regard to the natural sources of PAHs from terrestrial sources (non-anthropogenic burning of forests, woodlands, and moorlands due to lightning strikes, volcanic eruptions) and cosmic origin, their contributions have been estimated as being negligible to the overall emission of PAHs. However, it is rather difficult to make accurate estimations of PAH emissions because some PAHs are common to a number of these sources, and it is not easy to quantitatively determine how much of a particular PAH comes from a specific source [25]. Additionally, understanding the impact of particular emission sources on the different environmental compartments is crucial for proper risk assessment and risk management. PAHs are always emitted as a mixture, and the relative molecular concentration ratios are considered (often only as an assumption) to be characteristic of a given emission source. Table 2 lists typical diagnostic ratios taken from the literature that show the wide diversity of approaches [26].

Diagnostic ratios should be used with caution; the reactivity of some PAH species with other atmospheric species, such as ozone and/or oxides of nitrogen can change the diagnostic ratio [36-40]. The difference in chemical reactivity, volatility and solubility of PAH species may also introduce bias but to minimise this error, diagnostic ratios obtained from PAHs that have similar physico-chemical properties is mainly used [28]. Regardless, vehicular emissions were shown to be a major source of PAHs, whether the diagnostic ratios were coupled or not with principle component analysis, and irrespective of the region or season involved; moreover, studies indicated that diesel exhausts were the largest source of PAHs and nitro-PAHs, compared with the emissions of the other vehicles [39-42].

3. Diesel exhaust collection, extraction and chemical analysis

Diesel exhaust contains not only volatile species, which constitute the gaseous phase, but also solid agglomerates of spherical primary particles of approximately 15-30 nm diameter. These agglomerates are larger particles, in the range 60-200 nm (with a lognormal size distribution), which are produced by an accumulation mode corresponding to a coagulation of the primary particles. The agglomerates are made of solids (elemental carbon and metallic ashes) mixed with condensates and adsorbed material (including organic hydrocarbons). The accumulation mode can be accompanied by a nucleation mode, which consists of smaller particles (which are called ultra-fine particles of 10-20 nm) where the organic material is condensed on primary inorganic nuclei (sulphuric acid) [43]. It has been demonstrated that the ultra-fine particles are more easily volatilised, depending on temperature and dilution conditions, because they are mostly composed of volatile condensates and contain little solid material [44]. Therefore, discriminating between the gaseous species and the chemical species present in the solid state in diesel exhausts remains difficult because their formation processes depend on the sampling train design, sampling location, temperature, humidity and dilution rates of exhausts, and also on after-treatment devices.

3.1. Exhaust sampling and collection for PAHs analysis

Numerous instruments are used to monitor and characterise particle-bound or gas-phase PAHs in diesel tailpipe emissions. To obtain reliable and reproducible measurements, it is necessary to properly design the sampling system and to choose adequate materials for trapping the chemical substances to be measured (filters, sorbents, etc.). The majority of experiments are performed on chassis dynamometers under predetermined engine load; a heated hose transfers the exhaust from the tailpipe to full flow exhaust dilution tunnels, and therefore, the exhaust gases are collected at low temperature after being mixed and homogenised with the ambient air (with moderate dilution ratios ranging from 6 to 14) in the dilution tunnel, which is connected to a constant volume sampling system (CVS technique) [45, 46]. Chassis dynamometers can also be equipped with dilution systems such as critical-flow Venturi dilution tunnels [47], ejection dilutors [48], or rotating disk diluters [49], but undiluted exhaust can also be directly collected through a sampling probe inserted inside the tailpipe, which is a less conventional method[50].

3.1.1. Collection of the gas phase

The supports used for trapping diesel gaseous exhaust are often the same as those used for air sampling; polyurethane foam plugs (PUF) of appropriate density can trap semivolatile contaminants in aerosols without creating excessive back pressure, and they are typically used to collect the atmospheric gaseous phase [51]; however, sorbents such as XAD-2 resins can also be used to trap the air gas fraction [52]. XAD resins are nonionic macroporous polystyrene-divinylbenzene beads with macroreticular porosity and high surface areas, which give them their adsorptive characteristics for non-polar hydrophobic compounds such as PAHs or their derivatives. Therefore, we can also find these materials for the collection of semi-volatile PAHs and nitro-PAHs in the gas phase of diesel exhaust, because they are inexpensive and easy to handle, store and transport. Different adsorbent resins, such as XAD-2 or XAD-16, coupled to PUF cartridges (making "sandwiches" cartridges) can be used to collect PAHs in the gaseous phase [53-54]. The gaseous PAHs can also be collected using an annular denuder coated with different sorbents: the first stage is coated with XAD-4 resin, the second stage collects the particulate matter and the third stage is composed of a "sandwich" of polyurethane foam plugs and XAD-4 resin, to assess the volatilisation of PAHs from particles [55].

As previously mentioned, the vast majority of diesel exhaust is sampled in a dilution tunnel at low temperatures, where the condensation of the gaseous phase on solid particles is dominant. Additionally, operating at dilution ratios of approximately 10 is far from perfect because typical atmospheric dilution ratios are between 500 and 1000. A consequence is that the nucleation mode is favoured in the dilution tunnels where the saturation ratios are higher than those during atmospheric dilution, and the gaseous species coming out of the tailpipe are generally impoverished. However, one can be interested in the analysis of the gaseous exhaust at high temperature before the appearance of nucleation or condensation. A sampling method that differs from conventional collection methods was developed for this purpose and consists of trapping PAHs in an aqueous solution by gas bubbling and not on a solid sorbent. Therefore, when a particulate filter is incorporated after the diesel engine, the original method for trapping vapour-phase PAHs present in the hot undiluted gaseous exhaust is to absorb them inside an aqueous solution containing various additives, including a cationic surfactant as a solubilising agent [56].

3.1.2. Collection of the particulate phase

Diesel particulate matter (PM) is frequently defined from the material collected on a filter at a temperature of 52°C (or less) after dilution of the diesel exhaust with air. Even if the experiments conducted inside a dilution tunnel do not represent the full range of atmospheric conditions, these sampling conditions are far more common.

With regard to determining particulate PAHs in atmospheric aerosols, many different types of filters can be used and compared for retaining them; filters made of quartz, glass fibres, Teflon coated and nylon can be used [57]. For diesel exhaust, collection systems are also generally composed of glass fibre filters to trap the particulate matter, *e.g.*, Pallflex systems. However Teflon-coated glass fibre filters are preferred over glass fibre filters, because they are more inert to catalysing chemical transformations and are less moisture-sensitive [46].

The sampling procedure using filters to collect particle bound PAHs provides no information about the size of the particulate matter; therefore, a high-flow cascade impactor can be used to replace single filters, which collects particles based on size [58]. Many cascade impactors can separate the particulate matter using different quartz filters placed in each stages into eight size ranges, from particles smaller than 0.4 μm to sizes in the range of 6.6-10.5 μm [21]. Aluminium filters that have silicon grease sprayed onto their surfaces can also be used in each stage of the impactors to prevent the bouncing of the particles during the collection; therefore, the distribution of PAHs based on the particulate sizes can be evaluated [59].

Finally, many studies have focused on particles directly collected inside of a diesel particulate filter (DPF) to understand the effects that an exhaust after-treatment device may have on the formation of particle-bound PAHs; once the DPF filter is weighed for total mass emission control, the particles can be blown-off and recovered for subsequent chemical analysis [60].

3.2. Extraction of PAHs from the trapping media

3.2.1. Extraction and purification of trapped gaseous PAHs

Vapour-phase PAHs and nitro-PAHs present in the trapped phase are often solvent-extracted using conventional techniques such as an ultrasonic bath or Soxhlet apparatus. A large number of polyurethane foams are Soxhlet extracted with 10% diethyl ether in hexane, as described in US EPA method TO-13.

XAD-2 resins are also often Soxhlet extracted over the course of 8 hours with 120 mL of methylene chloride [53], whereas mixtures of PAHs and nitro-PAHs can be better extracted from XAD-16 resins using 300 mL of methylene chloride:acetonitrile 3:1 (v:v) for 16 hours [21]; the extracts are then concentrated to 1-2 mL using rotary evaporators or under nitrogen flow, and they must be purified before analysis to eliminate matrix interferences. For the cleanup process, the concentrated solutions can be introduced into a silica column that contains anhydrous sodium sulphate to exclude water, and then eluted with 20 mL of methylene chloride:hexane 1:2 (to recover PAHs) and 30 mL of acetone:hexane 1:2 (to better recover nitro-PAHs). Finally, after purification, another concentration step by evaporation is necessary to enhance the concentration of solutes to be analysed. Each of these steps (desorption, evaporation, purification, and evaporation) is time consuming, and large volumes of organic solvents are consumed; additionally, the evaporation concentration steps can be critical for obtaining quantitative results for the most volatile PAHs.

Another method for concentrating and cleaning PAHs (in one step) after their absorption in an aqueous medium is the use of a solid phase extraction process (SPE), which consists of percolating the PAHs solubilising aqueous medium through a short column containing hydrophobic packing material and eluting them directly inside of the analytical apparatus without any losses. This method is less time and solvent consuming [56].

3.2.2. Extraction of PAHs from diesel particles

Among all of the solid environmental polluted matrices studied, such as soils, sediments or urban dusts, carbonaceous diesel particulate matter is known to require severe conditions to obtain good recovery yields of the higher molecular weight PAHs [61]. It was even suggested that among all of the natural environmental matrices, diesel PM was the most refractory [62]. Despite this observation, only a few studies have a reliable optimisation of the PAHs extraction step from the carbonaceous matrix.

3.2.2.1. Conventional solvent extraction methods

In a significant number of studies, PAHs are extracted from PM trapped in glass-fibre filters in ultrasonic baths; for example, Karavakalis et al. used methylene chloride to extract three times the PAHs, with a total of 80 mL of solvent [45]. Riddle et al. used a 1:1 mixture of methylene chloride:hexane for 15 minutes, three times [63]. Fernandes et al. mentioned that using an ultrasonic bath for 1 hour with 70 mL of hexane was not as efficient as the Soxhlet extraction of PM with methylene chloride or a 3:1 mixture of methylene chloride:methanol [64]. However, Soxhlet extractions, even if they are more efficient than ultrasonic extractions, are longer; it can take 18 h to 24 h to extract PAHs using Soxhlet extractions [59, 65]. In many cases, even a long Soxhlet extraction does not result in the complete extraction of high weight PAHs and nitrated PAHs from diesel soot. Figure 1 shows that increasing the number of Soxhlet extraction cycles, using a classical extraction solvent for PAHs, does not permit the extraction of strongly adsorbed PAHs and nitro-PAHs from diesel soot; however, the aliphatic hydrocarbons are quantitatively extracted. In this example, the studied soot was very poor in the soluble organic fraction (SOF) and the pollutants were strongly adsorbed onto the partially graphitic surface and not absorbed into the condensation layer around the carbonaceous core, which can explain the difficulty of extracting high weight PAHs using classical techniques and solvents. However, more and more engines equipped with DPF result in the production of soot with very few condensed SOF, where the PAHs can be very difficult to extract; therefore, it seems risky to neglect the optimisation of the extraction step.

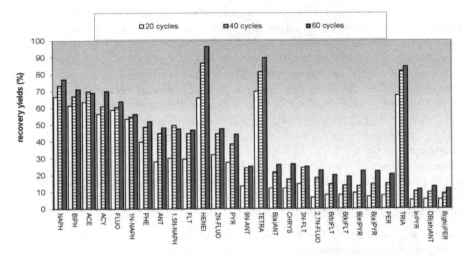

Figure 1. Soxhlet extraction of a mixture of spiked PAHs, nitro-PAHs and n-alkanes from 100 mg of diesel particulate matter, blown-off from a diesel particulate filter after an European NEDC driving cycle test. Extraction conditions: reflux of 120 mL methylene chloride, each cycle representing approximately 8 minutes.

Soxhlet extraction can be enhanced with hot Soxhlet, where heating is also applied to the extraction cavity (unlike conventional Soxhlet), but the temperature must be kept lower than the boiling point of the extracting solvent mixture to keep it in the liquid state; therefore, temperatures are not high, and even if the recoveries are slightly better than using conventional Soxhlet, it remains difficult to quantitatively extract high weight PAHs from poor SOF diesel particles [60].

Finally, before the analytical step, columns containing a mixture of silica gel and deactivated alumina can be used for the fractionation and purification of extracts; the elution with *n*-hexane permits the elimination of aliphatic hydrocarbons and a second elution using 25 mL of a 3:1 mixture of cyclohexane:methylene chloride permits the elution of PAHs and nitro-PAHs [40]. As previously mentioned, all of these extraction and cleanup steps are time and solvent consuming.

3.2.2.2. Recent rapid solvent extraction methods

Other extraction procedures, such as microwave-assisted extraction (MAE), supercritical fluid extraction (SFE) or assisted solvent extraction (ASE) can favourably replace Soxhlet or sonication extractions and yields cleaner extracts with minimal loss of volatile compounds and minimal use of solvents. Although these recent and more effective extraction methods are employed to extract PAHs or nitro-PAHs from solid matrices such as soils, sediments, plants or atmospheric dusts [66-70], they are only little used for the extraction of PAHs from collected diesel soot.

Microwave-assisted extraction uses closed inert vessels that contain the solid matrix and the organic solvents subjected to microwave irradiation (see Figure 2a). An important advantage of the MAE technique is the extraction rate acceleration due to microwave irradiation, resulting in an immediate heating to 120-140°C; therefore, extraction times on the order of a few minutes (approximately 30 minutes) can be obtained compared to a few hours when Soxhlet is used. Solvent volumes are low and range between 15 and 40 mL, depending on the quantity of matrix to be extracted. The solvent mixtures may have a dielectrical constant that is large enough to permit the heating transfer, and typical solvents such as acetonitrile or mixtures of toluene:acetone are employed to extract PAHs from atmospheric particles [71, 72]. In fact, the choice of the nature and the volume of the extracting solvents is especially important for the quantitative extraction of PAHs from diesel particles. Additionally, in the case of diesel soot with a very little SOF fraction, mixtures of drastic solvents such as pyridine:diethylamine or pyridine:acetic acid must be employed to quantitatively extract high weight PAHs or nitro-PAHs [73].

Supercritical fluid extraction exploits the unique properties of a supercritical fluid to more rapidly extract organic analytes from solid matrices, because the supercritical state imparts a great diffusion rate (like in the gaseous state) and a good solvating power (like in the liquid state). Carbon dioxide is the most used supercritical fluid because its critical temperature of 31°C and critical pressure of 74 bar are easily obtained. Supercritical fluids offer the opportunity to control the solvating power of the extraction fluid by varying the temperature (50 to 150 °C) and the pressure (100 to 400 bars), but also by adding minimal organic modifier (Figure 2b). In fact, the addition of small amounts of cosolvent into the CO_2 is absolutely necessary, not only to increase the PAH solubility in the extractant fluid but also to break the strong interactions between the aromatic retention sites of soot surface and the PAHs and nitro-PAHs: cosolvent mixtures of methylene chloride and toluene can be added to better extract PAHs from diesel soot [74], but other less conventional cosolvent mixtures, containing pyridine, are sometimes required [75]. SFE is an attractive method, as it leads to rapid extractions (less than 45 minutes, combining static and dynamic steps) and generates extracts ready for analysis without the need for additional concentration by means of solvent evaporation, because the extracted solutes are recovered into a very small volume of solvent. Moreover, SFE provides clean extracts because of its higher selectivity when compared to liquid solvent extraction techniques. Consequently, time consuming clean-up steps, which are also a source of analytical error, are not absolutely necessary after SFE extractions [76].

More recently, a method named accelerated solvent extraction (ASE) or pressurised fluid extraction (PFE) was developed. This method uses an organic solvent at a relatively high pressure and temperature to achieve more rapid extractions from solid matrices (Figure 2c) [77]. Elevated temperatures (150-200°C) permit to disrupt the strong solute-matrix interactions, and a high pressure (100-150 bars) forces the extraction solvent into the matrix pores. Less than 20 mL of solvent can be used to extract PAHs and nitro-PAHs from diesel PM; extractions with toluene [61], methylene chloride [16] or less conventional solvents based on pyridine [60] were performed.

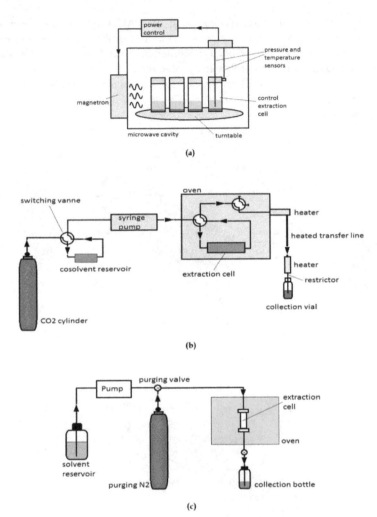

Figure 2. Extraction apparatus (a) Microwave-Assisted Extraction: MAE extraction vessels are placed on a turntable inside an oven and are subjected to microwave irradiations generated by a magnetron. A programmable microcomputer controls and monitors the power, temperature and pressure within the vessel (b) Supercritical Fluid Extraction: SFE incorporates pumps which produce the high pressures required for supercritical CO_2 work. An organic cosolvent can be added to the extraction fluid by a separate module. The extraction cell is placed in an oven which maintains a precise fluid temperature. A static mode (equilibration) and a dynamic mode (with a controlled flow-rate) can be performed successively. A restrictor valve (changing the CO_2 from condensed to gaseous phase) provides precise control over flow-rates (c) Assisted-Solvent Extraction: ASE operates by moving the extraction solvent through an extraction cell which is heated by direct contact of the oven. Extractions can be performed first in static mode; when the extraction is complete, compressed nitrogen moves the heated extracting solvent to the collection bottle.

Finally, we stress the importance of optimising the PAH extraction step from soot; a chemometric approach can be useful because the influence of several operating variables must be understood. Indeed, the use of an experimental design fully permits the evaluation of not only the effects of each variable but also the interaction effects between the studied parameters, and at last permits the construction of a mathematical model that relates the observed response (PAH recovery yields in %) to the various factors and to their combinations (Figure 3).

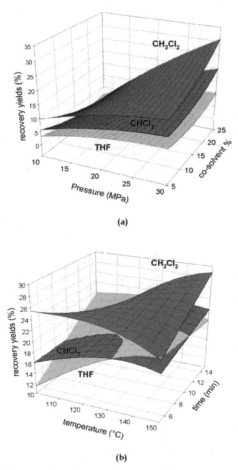

Figure 3. Modelisation of the extraction of benz[a]pyrene from 100 mg of spiked diesel soot, as a function of several influent factors (a) SFE recoveries as a function of pressure and percentage of co-solvent (methylene chloride, chloroform or tetrahydrofuran) added into supercritical CO_2, the temperature being fixed at 75°C and the extraction time at 10 minutes (static phase) and 20 minutes (dynamic phase) (b) ASE recoveries as a function of temperature, extraction time, and nature of the extraction solvent, the pressure being fixed at 100 bars.

3.2.2.3. *Extractions without solvent*

To avoid the use of organic solvents, other extraction techniques for chemical substances from particulate matter are proposed. Solid phase micro-extraction (SPME) seems to provide an alternative to solvent extraction methods [78]. Head-space SPME methods using polydimethylsiloxane fibres were tested on certified diesel particulate matter; it consisted in equilibrating PAHs sorbed on PM between a saline aqueous phase and the SPME fibre, and then the fibre was thermally desorbed inside the injection port of a gas chromatographer. However, this method was acceptable only for two to four PAHs congeners and was long (4 hours) [79]. Another method is to thermally desorb PAHs from filters or PM at 550°C in a thermogravimetric analyser, then recover them in a sampling bag where the vaporised phase is in equilibrium with a SPME fibre. Thereafter, PAHs are desorbed from the SPME fibre into a gas chromatographer [80]. However, for the quantitative analysis of high weight PAHs,the time necessary to establish the equilibrium (especially at higher concentration levels) is very long and can exceed 15h.

Direct thermal desorption (TD) can also be used for the desorption of PAHs from PM and this technique is generally hyphenated to a gas chromatographer for subsequent analysis. However, this fast extraction technique, which overcomes the main drawbacks of solvent extractions, is more often performed for PAHs sorbed on airborne particulate matter [81-83]; therefore, PAHs are desorbed from air dusts at 300-340°C (without pyrolytic degradation), cryogenically trapped and afterwards released in the gas chromatographer injection port at 325-400°C with a helium flow. Thermal desorption of PAHs from diesel exhaust particles seems to be more difficult and less reproducible, especially for high molecular weight PAHs [84].

Laser desorption-ionisation has also been studied to characterise pollutants immobilised on solid particles, and it offers the advantages of being orders of magnitude more sensitive for the analysis of PAHs and more rapid than solvent extraction. As will be discussed later, laser desorption systems are directly hyphenated to mass spectrometer analysers. Particulate PAHs and nitro-PAHs sorbed on soot particles are generally desorbed at an ionisation wavelength of 193 nm or 266 nm by an UV laser, mixed to a nebulised aerosol and transferred to the ion source region of the mass spectrometer [85]; solid samples to be desorbed can also be directly introduced with a probe into the ion source chamber [86, 87]. It has been described that this energetic UV laser desorption technology is vulnerable to matrix effects, and absolute quantification of PAHs seems to be difficult [88]. Therefore, new methods have appeared where particulate PAHs are deposited onto a probe, are irradiated with a less energetic pulsed Infra-Red laser beam (1064 nm) and the vaporised molecules are photoionised with UV radiation at 118 nm, so fragmentations are minimised and the sensitivity is higher [89].

3.3. Identification and quantitation of PAHs and their derivatives

3.3.1. *On-line mass spectrometry analysis*

To avoid time consuming sample preparation, we previously mentioned that PAHs can be desorbed from soot particles by IR or UV lasers, and after an ionisation step, they can be directly analysed by mass spectrometry. Generally, time-of-flight (TOF) mass spectrometers are used, and mass peaks ranging from 1 to 5000 g mol^{-1} are displayed. The advantages of this on-line analysis are that very small amounts of PM are required (as small as 10^{-12} to 10^{-11} g) and that the mass range is virtually unlimited [90]. Moreover, even if multiple mass spectra are required to yield a representative spectrum for each chemical species present in the particulate effluent, high acquisition rates allow a significant increase in time resolution and measurements can be performed within minutes [85]. The drawbacks are that only qualitative analyses are assured (identification of compound classes), due to effects of species variations in the small irradiated sample area and to the variability of the laser intensity [87]. Moreover, there is not any selectivity on PAHs isomers on mass spectra, and interferences due to alkylated compounds, carbon clusters, polyynes compounds and photofragments of nitro-PAHs are inevitable in spectra because of the complexity of diesel soot particles. On-line carbon speciation of diesel exhaust particles can also be performed by near-edge X-ray absorption spectroscopy (NEXAFS), but the same drawbacks can be underlined: only fingerprints can be established because PAHs isomers cannot be resolved and quantitative information cannot be proposed [91]. Consequently, a separation technique before the detection instrument seems essential to obtain isomer resolution and quantitation.

3.3.2. Chromatographic analysis

3.3.2.1. Gas chromatography (GC)

The vast majority of chromatographic separations of PAHs and nitro-PAHs in gas chromatography were obtained using classical columns with a length of 30 m, and an internal diameter of 0.25 mm; the open tube is generally coated with a stationary phase composed of cross-linked phenyl (5%) methyl (95%) siloxane (film thickness of 0.25 μm). However, many researchers have used a less polar stationary phase for the separation of PAHs (100% methyl), and a more polar one for nitro-PAHs (50% phenyl) [92]. Original stationary phases sometimes have to be employed to obtain an isomer resolution of non-priority high weight PAHs, which are nevertheless of considerable interest because they have a high carcinogenicity; smectic-liquid crystalline polysiloxane phases permitted the separation of these PAHs [93]. Another method for a better isomer resolution was to use a longer column, which was 50-60 m long [16]. The most used detector for detecting PAHs or nitro-PAHs after their chromatographic separation is the mass spectrometer, which has the advantages of being more sensitive and providing information for identification, compared to the flame ionisation detector [65]. Quadrupole mass analysers are the most used for PAHs detection, with electron ionization energies of 70 eV. Ion traps can also be used, which provide more accurate structural information (for isomer identification) in the MS-MS mode [63]. With regard to nitro-PAHs, the electron-capture detector (ECD) can be employed, which is a sensitive and selective detector [21]. MS detectors can also be used for the detection of nitro-PAHs, but with different ionisation modes than those applied for PAHs; the most frequent mode is negative chemical ionisation (NICI) with methane as a reagent gas [92], but electron monochromator mass spectrometry (EM-MS) appears to be more selective and specific [94].

It must be noted that the use of programmable temperature vaporisation (PTV) injection systems may enhance the sensitivity by injecting up to 40 μL of samples instead of the 1 μL used for classical split-splitless injectors [69]. However, GC injection systems can also be directly hyphenated to thermodesorption units, which permits on-line PAH thermodesorption, GC separation and MS detection [95], as mentioned in a previous chapter. Finally, it will be necessary to count on comprehensive two dimensional GC techniques (GCxGC) in the future to obtain increased separation power and better sensitivity [96].

3.3.2.2. High performance liquid chromatography (HPLC)

As PAHs are lipophilic compounds, separations are generally achieved using reversed phase liquid chromatography with an apolar octadecyl (C_{18}) column and a mobile phase composed of acetonitrile and water. However, phenyl-modified stationary phases can also be employed for the separation of PAHs and their oxidised derivatives, with methanol and water used as a mobile phase [97]. Even if Ultra-Violet detectors can be used to detect PAHs and nitroPAHs [98], fluorescence detectors are more frequently employed for PAHs because of their greater sensitivity and selectivity, even if one of the sixteen priority PAHs cannot be detected (acenaphtylene) [56]. Nitro-PAHs must be derivatised to be fluorescent (reduction to the amine), and other more appropriate detectors can be employed for their detection, such as chemiluminescence or coulometric detectors [17]. Mass spectrometers have also proven to be useful detectors; PAHs can be analysed using an atmospheric pressure chemical ionisation interface (APCI) between the separation column and a single-quadrupole detection system [97]; atmospheric pressure photoionisation interfaces (APPI) can also be coupled to triple quadrupole mass spectrometers, permitting the analysis of PAHs in the positive ion mode, and nitro-PAHs in the negative ion mode, but also providing structural information about the metabolites [99]. Analytical systems were developed to improve the sensitivity of nitro-PAHs analysis (over chemiluminescence detection) using a two dimensional HPLC procedure (on-line derivatisation and separation), an electrospray ionisation source (ESI) and a triple quadrupole mass spectrometer [100]. Finally, a highly original approach consisting of a hyphenated HPLC and GC-MS resulted in the better analysis of isomers of high weight PAHs from diesel particulate matter [101].

4. Conclusion

It is a known fact that polyaromatic hydrocarbons in diesel exhaust are harmful. Therefore, it seems crucial to characterise them in both the gaseous and particulate phases, even if they are not yet regulated pollutants. It has been shown here that a variety of measurement protocols are available and that, even if many regulations in diesel exhaust sampling exist, it is not the case for the whole analytical process. And even for the first step of the measurement, the dilution factor and sampling temperature range are not strictly specified, which can induce great variability in diesel particle mass and composition. For that reason, various conclusions have been drawn about the influence of fuel composition, after-treatment devices, engine load, etc., on PAHs emissions; however, results from different research groups remain difficult to compare and all of the analytical artefacts were not resolved. For example, differentiating between gaseous and particle-bound PAHs is still a problem, because of the desorption or re-adsorption of volatile compounds during sampling. Additionally, neglecting the vapour-phase analysis (as suggested in some studies) also introduces significant error, because the toxic equivalent factors of exhaust are not correctly evaluated. The second step of the analytical process is also, without a doubt, an important source of analytical bias. There is not a universal technique for the extraction of every possible PAH or nitro-PAH from a solid matrix, and all of the described techniques have advantages and disadvantages; for example, high time resolution when coupling the laser desorption directly with the detection is not compatible with isomeric resolution, and thermal desorption remains less reproducible than solvent extraction. With regard to solvent extraction, which is the most employed method for the extraction of PAHs and nitro-PAHs from diesel soot, its optimisation is regrettably widely neglected. Conventional extraction methods such as Soxhlet or ultrasonic extractions, which are generally investigated on diesel PM extractions, and conventional solvents, such as methylene chloride or hexane, are in many cases not sufficient to quantitatively extract high weight PAHs and nitro-PAHs from carbonaceous diesel PM [102]. It is particularly the case for "dry" diesel soot, which has a reduced portion of soluble organic fraction. Indeed, in this case, the sorption is dominated by strong surface adsorption while a high amount of SOF attenuates PAHs adsorption, blocking the energetic sites, and thus a simple phase partitioning dominates [103]. Consequently, the presence on the particles of relevant amounts of SOF helps the solvent extraction process; however, it is now relevant to notice that recent diesel engines that are equipped with oxidation catalysts and particulate filters produce leaner particulates, and as it previously mentioned, more drastic extraction conditions are required [61]. Therefore, in the future, it will be important to pay more attention on the optimisation of the extraction step to obtain quantitative results, especially for the highest weight and more toxic PAHs and nitro-PAHs. Then, new standard reference materials with low SOF, issued from more recent diesel engines, should be produced, characterised and commercialised to validate this important analytical step. Finally, it can be emphasised that studies on PAHs other than the sixteen priority ones, as well as on other oxygenated or sulphured derivatives, could be another important task in the future, considering the fact that some of them are particularly more dangerous. For those studies, comprehensive chromatographic techniques (GCxGC) could help in enhancing the resolution of hundreds of aromatic compounds that can be found on diesel particles.

Author details

F. Portet-Koltalo *and N. Machour
UMR CNRS 6014 COBRA, Université de Rouen, Evreux, France

Abbreviations and symbols

APCI   Atmospheric Pressure Chemical ionisation
APPI   Atmospheric Pressure PhotoIonisation

---

* Corresponding Author

ASE	Assisted Solvent Extraction
CVS	Constant Volume Sampling
DPF	Diesel Particulate Filter
ECD	Electron Capture Detector
EM-MS	Electron Monochromator Mass Spectrometry
ESI	Electro Spray Ionisation
GC	Gas Chromatography
HPLC	High Pressure Liquid Chromatography
IARC	International Agency for Research on Cancer Classification
K_H	Henry's law constant
Kow	n-octanol/water partition coefficient
MAE	Microwave Assisted Extraction
MS	Mass Spectrometry
NEDC	New European Driving Cycle
NEXAFS	Near Edge Xray Absorption Spectroscopy
NICI	Negative Chemical Ionisation
PAH/PAHs	Polycyclic Aromatic Hydrocarbon(s)
PM	Particulate Matter
PTV	Programmable Temperature Vaporisation
PUF	PolyUrethane Foam
SFE	Supercritical Fluid Extraction
SOF	Soluble Organic Fraction
SPE	Solid Phase Extraction
SPME	Solid Phase Micro Extraction
TD	Thermal Desorption
TOF	Time of Flight
US-EPA	United States Environmental Protection Agency
UV	Ultra Violet
WHO	World Health Organization
XAD/XAD-2/XAD-16	Resins with different macroreticular porosity

5. References

[1] Li Q, Wyatt A, Kamens R.M (2009) Oxidant generation and toxicity enhancement of aged-diesel exhaust. Atmospheric Environment. 43: 1037-1042.

[2] T.T.T, Lee B.-K (2009) Characteristics, toxicity, and source apportionment of polycylic aromatic hydrocarbons (PAHs) in road dust of Ulsan, Korea. Chemosphere. 74: 1245-1253.

[3] Andersson H, Piras E, Demma J, Hellman B, Brittebo E (2009) Low levels of the air pollutant 1-nitropyrene induce DNA damage, increased levels of reactive oxygen species and endoplasmic reticulum stress in human endothelial cells. Toxicology. 262: 57-64.

[4] Okuda T, Okamoto K, Tanaka S, Shen Z, Han Y, Huo Z (2010) Measurement and source identification of polycyclic aromatic hydrocarbons (PAHs) in the aerosol in Xi'an,

China, by using automated column chromatography and applying positive matrix factorization (PMF). Science of The Total Environment. 408: 1909-1914.

[5]   Duran A, Carmona M, Monteagudo J (2004) Modelling soot and SOF emissions from a diesel engine. Chemosphere. 56: 209-225.

[6]   Tan P, Hu Z, Deng K, Lu J, Lou D, Wan G (2007) Particulate matter emission modelling based on soot and SOF from direct injection diesel engines. Energy Conversion and Management. 48: 510-518.

[7]   Kittelson D, Watts W, Johnson J, Schauer J, Lawson D (2006) On-road and laboratory evaluation of combustion aerosols-Part 2: Summary of spark ignition engine results. Journal of Aerosol Science. 37: 931-949.

[8]   IARC (2010) IARC monographs on the evaluation of carcinogenic risks to humans, volume 92. Some non-heterocyclic polycyclic aromatic hydrocarbons and some related exposures.

[9]   Environmental Health Criteria EHC n°229 (2003) Selected nitro- and nitro oxy-polycyclic aromatic hydrocarbons. First draft prepared by Kielhorn J, Wahnschaffe U, Mangelsdorf U. Fraunhofer Institute of Toxicology and Aerosol Research, Hanover, Germany.

[10] EFSA (2008) Scientific Opinion of the Panel on Contaminants in the Food Chain on a request from the European Commission on Polycyclic Aromatic Hydrocarbons in Food. *The EFSA Journal* 724: 1-114.

[11] Third International Symposium on Chemistry and Biology, Michigan (1979) Polynuclear aromatic hydrocarbons, in: Jones P., Leber P. (Eds.), Third International Symposium on Chemistry and Biology: carcinogenesis and mutagenesis, Ann Arbor Science.

[12] U.S. EPA, North Carolina, USA (1990) United States Environmental Protection Agency, Cancer risk from outdoor exposure to air toxics, Vol.1, Final Report.

[13] IARC (1998) International Agency for Research on Cancer, list of IARC Evaluations, polynuclear aromatic hydrocarbons (PAH, Air quality guidelines for Europe). Copenhagen, WHO Regional Office for Europe, 105–117.

[14] Durant J.L, Busby Jr. W.F, Lafleur A.L, Penman B.W, Crespi C.L (1996) Human cell mutagenicity of oxygenated, nitrated and unsubstituted polycyclic aromatic hydrocarbons associated with urban aerosols. Mutation Research. 371: 3–4.

[15] Traversi D, Schilirò T, Degan R, Pignata C, Alessandria ., Gilli G (2011) Involvement of nitro-compounds in the mutagenicity of urban PM2.5 and PM10 in Turin. Mutation Research. 726: 54– 59.

[16] Bamford H.A, Bezabeh D.Z, Schantz M.M, Wise S.A, Baker J.E (2003) Determination and comparison of nitrated- polycyclic aromatic hydrocarbons measured in air and diesel particulate reference materials, Chemosphere. 50: 575–587.

[17] Yang X, Igarashi K, Tang N, Lin J.M, Wang W, Kameda T, Toriba A, Hayakawa K (2010) Indirect- and direct-acting mutagenicity of diesel, coal and wood burning-derived particulates and contribution of polycyclic aromatic hydrocarbons and nitro-polycyclic aromatic hydrocarbons. Mutation Research. 695: 29–34.

[18] Taga R, Tang N, Hattori T, Tamura K, Sakaic S,Toriba ., Kizu R, Hayakawa K (2005) Direct-acting mutagenicity of extracts of coal burning-derived particulates and contribution of nitro-polycyclic aromatic hydrocarbons. Mutation Research. 581: 91–95.

[19] Maricq M.M (2007) Chemical characterization of particulate emissions from diesel engines: a review. Aerosol Science. 38: 1079-1118.

[20] Chang C-T, Chen B-Y (2008) Toxicity assessment of volatile organic compounds and polycyclic aromatic hydrocarbons in motorcycle exhaust. Journal of Hazardous Materials. 153: 1262–1269.

[21] Chiang H-L, Lai Y-M, Chang S-Y (2012) Pollutant constituents of exhaust emitted from light-duty diesel vehicles. Atmospheric Environment. 47: 399-406.

[22] Turrio-Baldassarri L, Battistelli C.L, Conti L, Crebelli R, De Berardis B, Iamiceli A.L, Gambino M, Iannaccone S (2004) Emission comparison of urban bus engine fueled with diesel oil and 'biodiesel' blend. Science of the Total Environment. 327: 147–162.

[23] Bergvall C, Westerholm R (2009) Determination of highly carcinogenic dibenzopyrene isomers in particulate emissions from two diesel and two gasoline-fuelled light-duty vehicles. Atmospheric Environment. 43: 3883–3890.

[24] Ravindra K, Ranjeet S, Van Grieken R (2008) Review: Atmospheric polycyclic aromatic hydrocarbons: source attribution,emissions factors and regulation. Atmospheric Environment. 42: 2895-2921.

[25] Mostert M.M.R, Ayoko G.A, Kokot S (2010) Application of chemometrics to analysis of soil pollutants. Trends in Analytical Chemistry. 29: 430-435.

[26] Tobiszewski M, Namiesnik J (2012) Review: PAH diagnostic ratios for the identification of pollution emission Sources. Environmental Pollution. 162: 110-119.

[27] Zhang X.L, Tao S, Liu W.X, Yang Y, Zuo Q, Liu S.Z (2005) Source diagnostics of polycyclic aromatic hydrocarbons based on species ratios: a multimedia approach. Environmental Science and Technology. 39: 9109-9114.

[28] Ravindra K, Wauters E, Van Grieken R (2008) Variation in particulate PAHs levels and their relation with the trans-boundary movement of the air masses. Science of the Total Environment. 396: 100-110.

[29] Pies C, Hoffmann B, Petrowsky J, Yang Y, Ternes T.A, Hofmann T (2008) Characterization and source identification of polycyclic aromatic hydrocarbons (PAHs) in river bank soils. Chemosphere. 72: 1594-1601.

[30] De La Torre-Roche R.J, Lee W-Y, Campos-Díaz S.I (2009) Soil-borne polycyclic aromatic hydrocarbons in El Paso, Texas: analysis of a potential problem in the United States/Mexico border region. Journal of Hazardous Materials. 163: 946-958.

[31] Akyüz M, Çabuk H (2010) Gas particle partitioning and seasonal variation of polycyclic aromatic hydrocarbons in the atmosphere of Zonguldak, Turkey. Science of the Total Environment. 408: 5550-5558.

[32] Yunker M.B, Macdonald R.W, Vingarzan R, Mitchell R.H, Goyette D, Sylvestre S (2002) PAHs in the Fraser River basin: a critical appraisal of PAH ratios as indicators of PAH source and composition. Organic Geochemistry. 33: 489-515.

[33] Oliveira C, Martins N, Tavares J, Pio C, Cerqueira M, Matos M, Silva H, Oliveira C, Camoes F (2011) Size distribution of polycyclic aromatic hydrocarbons in a roadway tunnel in Lisbon, Portugal. Chemosphere. 83:1588-1596.

[34] Opuene K, Agbozu I.E, Adegboro O.O (2009) A critical appraisal of PAH indices as indicators of PAH source and composition in Elelenwo Creek, southern Nigeria. Environmentalist. 29: 47-55.

[35] Callén M.S, de la Cruz M.T, López J.M, Mastral A.M (2011) PAH in airborne particulate matter. Carcinogenic character of PM10 samples and assessment of the energy generation impact. Fuel Processing and. 92: 176-182.

[36] Katsoyiannis A, Terzi E, Cai Q-Y (2007) On the use of PAH molecular diagnostic ratios in sewage sludge for the understanding of the PAH sources. Is this use appropriate? Chemosphere. 69: 1337-1339.

[37] Galarneau E (2008) Source specificity and atmospheric processing of airborne PAHs: implications for source apportionment. Atmospheric Environment. 42: 8139-8149.

[38] Zhang W, Zhang S, Wan C, Yue D, Ye Y, Wang X (2008) Source diagnostics of polycyclic aromatic hydrocarbons in urban road runoff, dust, rain and canopy through fall. Environmental Pollution. 153: 594-601.

[39] Saarnio K, Sillanpää M, Hillamo R, Sandell E, Pennanen A.R, Solanen R.O (2008) Polycyclic aromatic hydrocarbons in size-segregated particular matter from six urban sites in Europe. Atmospheric Environment. 42: 9087-9097.

[40] Karavalakis G, Fontaras G, Ampatzoglou D, Kousoulidou M, Stournas S, Samaras Z, Bakeas E (2010) Effects of low concentration biodiesel blends application on modern passenger cars. Part 3: Impact on PAH, nitro PAH and oxy-PAH emissions. Environmental Pollution. 158: 1584-1594.

[41] Mari M, Harrison R.M, Schuhmacher M, Domingo J.L, Pongpiachan S (2010) Interferences over the sources and processes affecting polycyclic aromatic hydrocarbons in the atmosphere derived from measured data. Science of the Total Environment. 408: 2387-2393.

[42] Magara-Gomez K, Olson M.R, Okuda T, Walz K.A (2012) Sensitivity of hazardous air pollutant emissions to the combustion of blends of petroleum diesel and biodiesel fuel. Atmospheric Environment. 50: 307-313.

[43] Burtscher H (2005) Physical characterization of particulate emissions from diesel engines: a review. Aerosol Science. 36: 896-932.

[44] Bukowiecki N, Kittelson D, Watts W, Burtscher H, Weingartner E, Baltensperger U (2002) Real-time characterization of ultrafine and accumulation mode particles in ambient combustion aerosols. Journal of Aerosol Science. 33: 1139-1154.

[45] Karavalakis G, Bakeas E, Fontaras G, Stournas S (2011) Effect of biodiesel origin on regulated and particle-bound PAH (polycyclic aromatic hydrocarbon) emissions from a Euro 4 passenger car. Energy. 36: 5328-5337.

[46] Pakbin P, Ning Z, Schauer J.J, Sioutas C (2009) Characterization of Particle Bound Organic Carbon from Diesel Vehicles Equipped with Advanced Emission Control Technologies. Environmental Science & Technology. 43: 4679-4686.

[47] Liu Z.G, Berg D.R, Swor T.A, Schauer J.J (2008) Comparative Analysis on the Effects of Diesel Particulate Filter and Selective Catalytic Reduction Systems on a Wide Spectrum of Chemical Species Emissions. Environmental Science & Technology. 42: 6080-6085.

[48] He C, Ge Y, Tan J, You K, Han X, Wang J (2010) Characteristics of polycyclic aromatic hydrocarbons emissions of diesel engine fueled with biodiesel and diesel. Fuel. 89: 2040-2046.

[49] Young L-H, Liou Y-J; Cheng M-T, Lu J-H, Yang H-H, Tsai Y.I, Wang L-C, Chen C-B, Lai J-S (2012) Effects of biodiesel, engine load and diesel particulate filter on nonvolatile particle number size distributions in heavy-duty diesel engine exhaust. Journal of Hazardous Materials. 200, 282-289.

[50] Heeb N.V, Schmid P, Kohler M, Gujer E, Zennegg M, Wenger D, Wichser A, Ulrich A, Gfeller U, Honegger P, Zeyer K, Emmenegger L, Petermann J-L, Czerwinski J, Mosimann T, Kasper M, Mayer A (2008) Secondary Effects of Catalytic Diesel

Particulate Filters: Conversion of PAHs versus Formation of Nitro-PAHs. Environmental Science & Technology. 42: 3773-3779.

[51] Dimashki M, Harrad S, Harrison R.M (2000) Measurements of nitro-PAH in the atmospheres of two cities. Atmospheric Environment. 34: 2459-2469.

[52] Esen F, Tasdemir Y, Vardar N (2008) Atmospheric concentrations of PAHs, their possible sources and gas-to-particle partitioning at a residential site of Bursa, Turkey. Atmospheric Research. 88: 243-255.

[53] de Abrantes R, de Assunçao J.V, Pesquero C.R (2004) Emission of polycyclic aromatic hydrocarbons from light-duty diesel vehicles exhaust. Atmos. Environ. 38: 1631-1640.

[54] Yang H-H, Hsieh L-T, Liu H-C, Mi H-H (2005) Polycyclic aromatic hydrocarbon emissions from motorcycles. Atmospheric Environment. 39: 17-25.

[55] Zielinska B, Sagebiel J, Arnott W.P, Rogers C.F, Kelly K.E, Wagner D.A, Lighty J.S, Sarofim A.F, Palmer G (2004) Phase and Size Distribution of Polycyclic Aromatic Hydrocarbons in Diesel and Gasoline Vehicle Emissions. Environmental Science & Technology. 38: 2557-2567.

[56] Portet-Koltalo F, Preterre D, Dionnet F (2011) A new analytical methodology for a fast evaluation of semi-volatile polycyclic aromatic hydrocarbons in the vapor phase downstream of a diesel engine particulate filter. Journal of Chromatography A. 1218: 981-989.

[57] Borras E, Tortajada-Genaro L.A (2007) Characterization of polycyclic aromatic hydrocarbons in atmospheric aerosols by gas-chromatography-mass spectrometry. Analytica Chimica Acta. 583: 266-276.

[58] Song W.W, He K.B, Wang J.X, Wang X.T, Shi X.Y, Yu C, Chen W.M, Zheng L (2011) Emissions of EC, OC, and PAHs from Cottonseed Oil Biodiesel in a Heavy-Duty Diesel Engine. Environmental Science & Technology. 45: 6683-6689.

[59] Lin Y-C, Tsai C-H, Yang C-R, Wu C.J, Wu T-Y, Chang-Chien G-P (2008) Effects on aerosol size distribution of polycyclic aromatic hydrocarbons from the heavy-duty diesel generator fueled with feedstock palm-biodiesel blends. Atmospheric Environment. 42: 6679-6688.

[60] Oukebdane K, Portet-Koltalo F, Machour N, Dionnet F, Desbène P.L (2010) Comparison of hot Soxhlet and accelerated solvent extractions with microwave and supercritical fluid extractions for the determination of polycyclic aromatic hydrocarbons and nitrated derivatives strongly adsorbed on soot collected inside a diesel particulate filter. Talanta. 82: 227-236.

[61] Turrio-Baldassarri L, Battistelli C.L, Iamiceli A.L (2003) Evaluation of the efficiency of extraction of PAHs from diesel particulate matter with pressurized solvents. Analytical and Bioanalytical Chemistry. 375: 589-595.

[62] Benner B.A (1998) Summarizing the Effectiveness of Supercritical Fluid Extraction of Polycyclic Aromatic Hydrocarbons from Natural Matrix Environmental Samples. Analytical Chemistry. 70: 4594-4601.

[63] Riddle S.G, Jakober C.A, Robert M.A, Cahill T.M, Charles M.J, Kleeman M.J (2007) Large PAHs detected in fine particulate matter emitted from light-duty gasoline vehicles. Atmospheric Environment. 41: 8658-8668.

[64] Fernandes M.B, Brooks P (2003) Characterization of carbonaceous combustion residues: II. Nonpolar organic compounds. Chemosphere. 53: 447-458.

[65] Rojas N.Y, Milquez H.A, Sarmiento H (2011) Characterizing priority polycyclic aromatic hydrocarbons (PAH) in particulate matter from diesel and palm oil-based biodiesel B15 combustion. Atmospheric Environment. 45: 6158-6162.

[66] Hawthorne S.B, Grabanski C.B, Martin E, Miller D.J (2000) Comparisons of Soxhlet extraction, pressurized liquid extraction, supercritical fluid extraction and subcritical water extraction for environmental solids: recovery, selectivity and effects on sample matrix. Journal of Chromatography A. 892: 421-433.

[67] Priego-Capote F, Luque-Garcia J, Luque de Castro M (2003) Automated fast extraction of nitrated polycyclic aromatic hydrocarbons from soil by focused microwave-assisted Soxhlet extraction prior to gas chromatography-electron-capture detection. Journal of Chromatography A. 994: 159-167.

[68] Yusa V, Quintas G, Pardo O, Pastor A, Guardia M (2006) Determination of PAHs in airborne particles by accelerated solvent extraction and large-volume injection-gas chromatography-mass spectrometry. Talanta. 69: 807-815.

[69] Bruno P, Caselli M, de Gennaro G, Tutino M (2007) Determination of polycyclic aromatic hydrocarbons (PAHs) in particulate matter collected with low volume samplers. Talanta. 72: 1357-1361.

[70] Itoh N, Fushimi A, Yarita T, Aoyagi Y, Numata M (2011) Accurate quantification of polycyclic aromatic hydrocarbons in dust samples using microwave-assisted solvent extraction combined with isotope-dilution mass spectrometry. Analytica Chimica Acta. 699: 49-56.

[71] Shu Y.Y, Tey S.Y, Wu D.K (2003) Analysis of polycyclic aromatic hydrocarbons in airborne particles using open-vessel focused microwave-assisted extraction. Analytica Chimica Acta. 495: 99-108.

[72] Castro D, Slezakova K, Delerue-Matos C, Alvim-Ferraz M.C, Morais S, Pereira MC (2011) Polycyclic aromatic hydrocarbons in gas and particulate phases of indoor environments influenced by tobacco smoke: Levels, phase distributions, and health risks. Atmospheric Environment. 45: 1799-1808.

[73] Portet-Koltalo F, Oukebdane K, Dionnet F, Desbène P.L (2008) Optimisation of the extraction of polycyclic aromatic hydrocarbons and their nitrated derivatives from diesel particulate matter using microwave-assisted extraction. Analytical and Bioanalytical Chemistry. 390: 389-398.

[74] Jones C, Chughtai A, Murugaverl B, Smith D (2004) Effects of air/fuel combustion ratio on the polycyclic aromatic hydrocarbon content of carbonaceous soots from selected fuels. Carbon. 42: 2471-2484.

[75] Portet-Koltalo F, Oukebdane K, Dionnet F, Desbène P.L (2009) Optimisation of supercritical fluid extraction of polycyclic aromatic hydrocarbons and their nitrated derivatives adsorbed on highly sorptive diesel particulate matter. Analytica Chimica Acta. 651: 48-56.

[76] Bowadt S, Hawthorne S.B (1995) Supercritical fluid extraction in environmental analysis. Journal of Chromatography A. 703: 549-571.

[77] Ryno M, Rantanen L, Papaioannou E, Konstandopoulos A.G, Koskentalo T, Kavela K (2006) Comparison of pressurized fluid extraction, Soxhlet extraction and sonication for the determination of polycyclic aromatic hydrocarbons in urban air and diesel exhaust particulate matter. Journal of Environmental Monitoring. 8: 488-493.

[78] Tang B, Isacsson U (2008) Analysis of Mono- and Polycyclic Aromatic Hydrocarbons Using Solid-Phase Microextraction: State-of-the-Art. Energy & Fuels. 22: 1425-1438.

[79] Vaz J.M (2003) Screening direct analysis of PAHS in atmospheric particulate matter with SPME. Talanta. 60: 687-693.

[80] Ballesteros R, Hernandez J, Lyons L (2009) Determination of PAHs in diesel particulate matter using thermal extraction and solid phase micro-extraction. Atmospheric Environment. 43: 655-662.

[81] Van Drooge B.L, Nikolova I, Ballesta P.P (2009) Thermal desorption gas chromatography-mass spectrometry as an enhanced method for the quantification of polycyclic aromatic hydrocarbons from ambient air particulate matter. Journal of Chromatography A. 1216: 4030-4039.

[82] Gil-Molto J, Varea M, Galindo N, Crespo J (2009) Application of an automatic thermal desorption-gas chromatography-mass spectrometry system for the analysis of polycyclic aromatic hydrocarbons in airborne particulate matter. Journal of Chromatography A. 1216: 1285-1289.

[83] Ancelet T, Davy P.K, Trompetter W.J, Markwitz A, Weatherburn D.C (2011) Carbonaceous aerosols in an urban tunnel. Atmospheric Environment. 45: 4463-4469.

[84] Lavrich R.J, Hays M.D (2007) Validation Studies of Thermal Extraction-GC/MS Applied to Source Emissions Aerosols. 1. Semivolatile Analyte - Nonvolatile Matrix Interactions. Analytical Chemistry. 79: 3635-3645.

[85] Kalberer M, Morrical B.D, Sax M, Zenobi R (2002) Picogram Quantitation of Polycyclic Aromatic Hydrocarbons Adsorbed on Aerosol Particles by Two-Step Laser Mass Spectrometry. Analytical Chemistry. 74: 3492-3497.

[86] Dotter R.N, Smith C.H, Young M.K, Kelly P.B, Jones A.D, McCauley E.M, Chang D.P.Y (1996) Laser Desorption/Ionization Time-of-Flight Mass Spectrometry of Nitrated Polycyclic Aromatic Hydrocarbons. Analytical Chemistry. 68: 2319-2324.

[87] Dobbins R.A, Fletcher R.A, Benner B.A, Hoeft S (2006) Polycyclic aromatic hydrocarbons in flames, in diesel fuels, and in diesel emissions. Combustion and Flame. 144: 773-781.

[88] Bente M, Sklorz M, Streibel T, Zimmermann R (2008) Online Laser Desorption-Multiphoton Postionization Mass Spectrometry of Individual Aerosol Particles: Molecular Source Indicators for Particles Emitted from Different Traffic-Related and Wood Combustion Sources. Analytical Chemistry. 80: 8991-9004.

[89] Ferge T, Muhlberger F, Zimmermann R (2005) Application of Infrared Laser Desorption Vacuum-UV Single-Photon Ionization Mass Spectrometry for Analysis of Organic Compounds from Particulate Matter Filter Samples. Analytical Chemistry. 77: 4528-4538.

[90] Faccinetto A, Desgroux P, Ziskind M, Therssen E, Focsa C (2011) High-sensitivity detection of polycyclic aromatic hydrocarbons adsorbed onto soot particles using laser desorption/laser ionization/time-of-flight mass spectrometry: An approach to studying the soot inception process in low-pressure flames. Combustion and Flame. 158: 227-239.

[91] Braun A, Mun B.S, Huggins F.E, Huffman G.P (2006) Carbon Speciation of Diesel Exhaust and Urban Particulate Matter NIST Standard Reference Materials with C(1s) NEXAFS Spectroscopy. Environmental Science & Technology. 41: 173-178.

[92] Kawanaka Y, Sakamoto K, Wang N, Yun S-J (2007) Simple and sensitive method for determination of nitrated polycyclic aromatic hydrocarbons in diesel exhaust particles

by gas chromatography-negative ion chemical ionisation tandem mass spectrometry. Journal of Chromatography A. 1163: 312-317.

[93] Schubert P, Schantz M.M, Sander L.C, Wise S.A (2003) Determination of Polycyclic Aromatic Hydrocarbons with Molecular Weight 300 and 302 in Environmental-Matrix Standard Reference Materials by Gas Chromatography/Mass Spectrometry. Analytical Chemistry. 75: 234-246.

[94] Ratcliff M.A, Dane A.J, Williams A, Ireland J, Luecke J, McCormick R.L, Voorhees K.J (2010) Diesel Particle Filter and Fuel Effects on Heavy-Duty Diesel Engine Emissions. Environmental Science & Technology. 44: 8343-8349.

[95] Bente M, Sklorz M, Streibel T, Zimmermann R (2009) Thermal Desorption-Multiphoton Ionization Time-of-Flight Mass Spectrometry of Individual Aerosol Particles: A Simplified Approach for Online Single-Particle Analysis of Polycyclic Aromatic Hydrocarbons and Their Derivatives. Analytical Chemistry. 81: 2525-2536.

[96] Ozel M.Z, Hamilton J.F, Lewis A.C (2011) New Sensitive and Quantitative Analysis Method for Organic Nitrogen Compounds in Urban Aerosol Samples. Environmental Science & Technology. 45: 1497-1505.

[97] Letzel T, Poschl U, Wissiack R, Rosenberg E, Grasserbauer M, Niessner R (2001) Phenyl-Modified Reversed-Phase Liquid Chromatography Coupled to Atmospheric Pressure Chemical Ionization Mass Spectrometry: A Universal Method for the Analysis of Partially Oxidized Aromatic Hydrocarbons. Analytical Chemistry. 73: 1634-1645.

[98] El Haddad I, Marchand N, Dron J, Temime-Roussel B, Quivet E, Wortham H, Jaffrezo J.L, Baduel C, Voisin D, Besombes J.L, Gille G(2009) Comprehensive primary particulate organic characterization of vehicular exhaust emissions in France. Atmospheric Environment. 43: 6190-6198.

[99] Hutzler C, Luch A, Filser J.G (2011) Analysis of carcinogenic polycyclic aromatic hydrocarbons in complex environmental mixtures by LC-APPI-MS/MS. Analytica Chimica Acta. 702: 218-224.

[100] Miller-Schulze J.P, Paulsen M, Toriba A, Hayakawa K, Simpson C.D (2007) Analysis of 1-nitropyrene in air particulate matter standard reference materials by using two-dimensional high performance liquid chromatography with online reduction and tandem mass spectrometry detection. Journal of Chromatography A. 1167: 154-160.

[101] Bergvall C, Westerholm R (2008) Determination of 252-302 Da and tentative identification of 316-376 Da polycyclic aromatic hydrocarbons in Standard Reference Materials 1649a Urban Dust and 1650b and 2975 Diesel Particulate Matter by accelerated solvent extraction-HPLC-GC-MS. Analytical and Bioanalytical Chemistry. 391: 2235-2248.

[102] Jonker M.T.O, Koelmans A.A (2002) Extraction of Polycyclic Aromatic Hydrocarbons from Soot and Sediment: Solvent Evaluation and Implications for Sorption Mechanism. Environmental Science & Technology. 36: 4107-4113.

[103] Endo S, Grathwohl P, Haderlein S.B, Schmidt T.C (2009) Effects of Native Organic Material and Water on Sorption Properties of Reference Diesel Soot. Environmental Science & Technology. 43: 3187-3193.

Chapter 4

Combustion and Exhaust Emission Characteristics of Diesel Micro-Pilot Ignited Dual-Fuel Engine

Ulugbek Azimov, Eiji Tomita and Nobuyuki Kawahara

Additional information is available at the end of the chapter

1. Introduction

To satisfy increasingly strict emissions regulations, engines with alternative gaseous fuels are now widely used. Natural gas and synthesis gas appear to be greener alternatives for internal combustion engines [1-3]. In many situations where the price of petroleum fuels is high or where supplies are unreliable, the syngas, for example, can provide an economically viable solution. Syngas is produced by gasifying a solid fuel feedstock such as coal or biomass. The biomass gasification means incomplete combustion of biomass resulting in production of combustible gases. Syngas consists of about 40% combustible gases, mainly carbon monoxide (CO), hydrogen (H_2) and methane (CH_4). The rest are non-combustible gases and consists mainly of nitrogen (N_2) and carbon dioxide (CO_2). Varying proportions of CO_2, H_2O, N_2, and CH_4 may be present [4].

H_2 as a main component of a syngas has very clean burning characteristics, a high flame propagation speed and wide flammability limits. H_2 has a laminar combustion speed about eight times greater than that of natural gas, providing a reduction of combustion duration and as a result, an increase in the efficiency of internal combustion (IC) engines, if the H_2 content in the gaseous fuel increases. Main point of interest in increasing H_2 content in the gaseous fuel is that with the addition of H_2, the lean limit of the gas operation can be extended, without going into the lean misfire region. Lean mixture combustion has a great potential to achieve higher thermal efficiency and lower emissions [5]. In particular, the lean mixture combustion will result in low and even extremely low NOx levels with only a slight increase in hydrocarbons [6, 7].

Some gas engines fueled with syngas have been developed recently [8-10]. Most of them have a spark-ignition (SI) combustion system. An SI engine is not suitable for this kind of fuel under high load conditions because of the difficulty in achieving stable combustion due to the fluctuation of the syngas components. In addition, the syngas is a low energy density fuel and the extent of power degrading is large when compared with high-energy density fuels like gasoline and natural gas. Natural gas and syngas have high auto-ignition temperature and hence cannot be used in CI engines without a means of initiating combustion, as the temperature attained at the end of the compression stroke is too low for the mixture to be auto-ignited. Therefore, dual-fuel-mode engine operation is required, in which gaseous fuel is ignited by pilot diesel fuel. Dual fueling can serve as a way of allowing the current fleet of CI engines to reduce their dependence on conventional diesel fuel while minimizing harmful emissions.

A number of researchers have performed experiments with natural gas and syngas to determine engine performance and exhaust emissions in dual-fuel engines. Their results indicate that lower NOx and smoke can be achieved in dual-fuel engines compared with conventional diesel engines, while maintaining the same thermal efficiency as a diesel engine. McTaggart-Cowan et al. [11] investigated the effect of high-pressure injection on a pilot-ignited, directly injected natural gas engine. They found that at high loads, higher injection pressures substantially reduce PM emissions. At low loads, the amount of PM emissions are independent of the injection pressure. Without EGR, NOx emissions are slightly increased at higher injection pressures due to the faster and more intense combustion caused by improved mixing of air and fuel and increased in-cylinder temperature. Su and Lin [12] studied the amount of pilot injection and the rich and lean boundaries of natural gas dual-fuel engines. They found that there is a critical amount of pilot diesel fuel for each load and speed. Tomita et al. [13] investigated the combustion characteristics and performance of the supercharged syngas with micro-pilot (injected fuel - 2 mg/cycle) ignition in a dual-fuel engine. They found that premixed flame of syngas-air mixture develops from multiple flame kernels produced by the ignition of diesel fuel. It was found that with the certain increase of hydrogen content in syngas the engine could operate even at equivalence ratio of 0.45 with stable combustion and high efficiency, because the increased hydrogen content enhanced the lean limit of the mixture. Liu and Karim [14] concluded that the observed values of the ignition delay in dual-fuel operation are strongly dependent on the type of gaseous fuels used and their concentrations in the cylinder charge. They showed that changes in the charge temperature during compression, preignition energy release, external heat transfer to the surroundings, and the contribution of residual gases appear to be the main factors responsible for controlling the length of the ignition delay of the engine.

The autoignition of the premixed mixture in the end-gas region is affected by the composition of syngas, in particular, by the amount of H_2, CO, CO_2 and CH_4 in the gas. H_2 has low ignition energy, and therefore, is easier to ignite, that results in a stronger tendency to autoignition and knock. H_2 has a flame speed much greater than that of hydrocarbon fuels. Also, it has a lean limit of $\phi_{lim}=0.1$, much lower than the theoretical limit of methane ($\phi_{lim}=0.5$) [15]. Carbon dioxide, on the other hand, can weaken the reactivity of the in-cylinder mixture by diluting it, which results in a longer ignition delay time and slower heat release rate. Methane has excellent anti-knock properties, but suffers from low flame propagation rates and high auto-ignition temperature. Carbon monoxide can also affect the reactivity of the mixture. In fact, the oxidation of CO in the presence of H_2 is important question concerning the syngas oxidation mechanism. It is well known that the overall reactivity is greatly accelerated if trace amounts of H_2 and moisture are present. The oxidation route between CO and OH is the dominant pathway, and it accounts for a significant portion of the heat release [16].

The aforementioned results suggest that if certain operating conditions are maintained, including the control of pilot fuel injection quantity, pressure and timing, gaseous fuel equivalence ratio, and EGR rate, a compromise between increased efficiency and low exhaust emissions can be achieved. In this chapter, we document the range of operating conditions under which the new higher-efficiency PREMIER (**PRE**mixed **M**ixture **I**gnition in the **E**nd-gas **R**egion) combustion mode was experimentally tested. The objective of this work was in brief to discuss conventional micro-pilot injected dual-fuel combustion and in detail to explain about PREMIER combustion and emission characteristics in a pilot ignited supercharged dual-fuel engine fueled with natural gas and syngas, and to study the effect of H_2 and CO_2 content in syngas on the combustion and emission formation over the broad range of equivalence ratios under lean conditions.

2. Conventional micro-pilot injected dual-fuel combustion

2.1. Experimental procedure and conditions

This study used water-cooled four-stroke single-cylinder engine, with two intake and two exhaust valves, shown in Figure 1 (A). In this engine, the autoignition of a small quantity of diesel pilot fuel (2 mg/cycle), injected into the combustion chamber before top dead center, initiates the combustion. The burning diesel fuel then ignites the gaseous fuel. The pilot fuel was ultra low-sulfur (<10ppm) diesel. A commercial solenoid-type injector that is typically used for diesel-only operations was modified. A nozzle of the commercial injector with seven holes was replaced by the one with four holes of 0.1 mm in diameter to ensure a small quantity of injected fuel.

Diesel fuel injection timing and injection duration were controlled through the signals transferred to the injector from the injector driver. A common rail injection system (ECD U2-P, Denso Co.) was employed to supply the constant injection pressure to the injector. The common rail pressure was set and controlled via computer. The fuel injection pressure varied from 40 MPa to 150 MPa, and the injected pilot diesel fuel quantity was 2 mg/cycle and 3 mg/cycle. The experimental conditions and different types of primary gaseous fuel compositions used in this study are given in Table 1 and Table 2, respectively.

The in-cylinder pressure history of combustion cycles was measured with KISTLER-6052C pressure transducer in conjunction with a $0.5°$ crank-angle encoder to identify the piston location. The rate of heat release (ROHR) was calculated using this equation [17]:

$$\frac{dQ}{d\theta} = \frac{\gamma}{\gamma-1}p\frac{dV}{d\theta} + \frac{1}{\gamma-1}V\frac{dp}{d\theta} \tag{1}$$

where θ is the crank angle (CA), p is the in-cylinder pressure at a given crank angle, V is the cylinder volume at that point and γ is the specific heat ratio. The ROHR represents the rate of energy release from the combustion process. The combustion transition from the first stage (slow flame propagation) to the second stage (end-gas autoignition) is identified from the ROHR. CO and NOx emissions were measured with a four-component analyzer Horiba PG-240, smoke was measured with a smoke meter Horiba MEXA-600s and HC emissions were measured with Horiba MEXA-1170HFID.

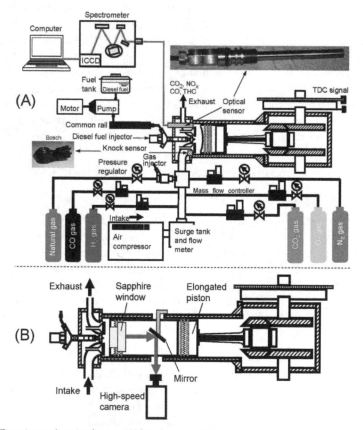

Figure 1. Experimental engine layout. (A) bench engine, (B) optical engine

An elongated cylinder liner and elongated piston were installed on the engine, Figure 1 (B), to visualize dual-fuel combustion events and to capture images of combustion in an optical engine. The engine mentioned above in Figure 1 (A) was modified to allow facilitating the visualization experiments.

Engine type	4-stroke, single cylinder, water cooled
Bore x Stroke	96x108 mm
Swept volume	781.7 cm3
Compression ratio	16
Combustion system	Dual-fuel, direct injection
Combustion chamber	Shallow dish
Engine speed	1000 rpm
Intake pressure	101 kPa, 200 kPa
Injection system	Common-rail
Pilot fuel injection pressure	
- *Natural gas case*	40 MPa, 80 MPa, 120 MPa, 150 MPa
- *Syngas case*	80 MPa
Pilot fuel injection quantity	
- *Natural gas case*	2 mg/cycle, 3 mg/cycle
- *Syngas case*	3 mg/cycle
Nozzle hole x diameter	
- *Natural gas case*	3x0.08 mm, 3x0.10 mm, 4x0.10 mm
- *Syngas case*	4x0.10 mm
Equivalence ratio	
- *Natural gas case*	0.6
- *Syngas case*	Variable
EGR rate	
- *Natural gas case*	10%, 20%, 30%, 40%, 50%
- *Syngas case*	none
EGR composition	N_2-86%, O_2-10%, CO_2-4%

Table 1. Experimental conditions

Gas type	Composition						
	H_2 (%)	CO (%)	CH_4 (%)	CO_2 (%)	N_2 (%)	LHV (MJ/kg)	Source
Type 1	13.7	22.3	1.9	16.8	45.3	4.13	BMG
Type 2	20.0	22.3	1.9	16.8	39.0	4.99	BMG
Type 3	56.8	22.3	1.9	16.8	2.2	13.64	COG
Type 4	13.7	22.3	1.9	23.0	39.1	3.98	BMG
Type 5	13.7	22.3	1.9	34.0	28.1	3.74	BMG
Type 6	56.8	5.9	29.5	2.2	5.6	38.69	COG
Type 7	56.8	29.5	5.9	2.2	5.6	20.67	COG
Type 8	100.0	-	-	-	-	119.93	Hydrogen
	CH_4 (%)	C_2H_6 (%)	C_3H_8 (%)	n-C_4H_{10} (%)		LHV (MJ/kg)	
Natural gas	88.0	6.0	4.0	2.0		49.20	

Table 2. Gas composition

A sapphire window was installed on the top of the elongated piston. Temporal and spatial evolutions of visible flames were investigated by acquiring several images per cycle with MEMRECAM fx-K5, a high speed digital camera with the frame rate of 8000 fps in combination with a $45°$ mirror located inside the elongated piston. An engine shaft encoder and delay generator were used to implement camera-engine synchronization. The engine was first operated with motoring, then it was fired on only one cycle during which combustion images were captured.

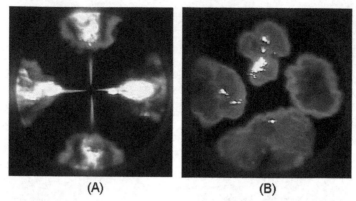

(A)   (B)

Figure 2. Dual fuel combustion of syngas with different amounts of injected diesel fuel (A) 10 mg and (B) 2 mg

Figure 2 shows the effect of pilot diesel fuel amount on combustion of syngas as a primary fuel. A distinct separation of diesel diffusion flame and syngas premixed flame is seen when larger amount of diesel fuel, 10 mg, is injected. The mixture burns only when the required oxygen is present. As a result, the flame speed is limited by the rate of diffusion. For syngas well-premixed mixture, the flame is not limited by the rate of diffusion and the mixture burning rate is much faster than that of diesel-air mixture. When only 2 mg of diesel fuel is injected the fuel evaporates and mixes with the oxidizer much faster providing distributed ignition centers in the cylinder for syngas-air mixture which is burnt as a premixed flame propagating towards the cylinder wall.

Figure 3 shows the dual-fuel combustion with natural gas (A) and syngas (B) as a primary fuel. The flame area growth for syngas is slower due to the presence of CO_2 and CO. Although syngas contains H_2, the effect of H_2 on flame propagation is not clearly seen due to simultaneous oxidation of H_2 and CO and the effect of CO_2 diluting the air-fuel mixture. The effect of CO oxidation in the presence of H_2 on flame propagation is a quite complex topic and is outside the scope of this chapter.

3. PREMIER micro-pilot ignited dual-fuel combustion

3.1. Concept of PREMIER combustion

Before giving a description to PREMIER (**PRE**mixed Mixture Ignition in the End-gas Region) combustion, it is necessary to explain the differences of phenomenological outline between conventional combustion and knocking combustion in the dual-fuel engine. Conventional combustion is a combustion process which is initiated by a timed pilot ignited fuel and in which the multiple flame fronts caused by multiple ignition centers of pilot fuel, moves completely across the combustion chamber in a uniform manner at a normal velocity. Knocking combustion is a combustion process in which some part or all of the charge may be consumed (autoignited) in the end-gas region at extremely high rates. Much evidence of end-gas mixture auto-ignition followed by knocking combustion can be obtained from high-speed laser shadowgraphs [18], high-speed Schlieren photography [19], chemiluminescent emission [20], and laser-induced fluorescence [21]. In addition, Stiebels *et al.* [22] and Pan and Sheppard [23] showed that multiple autoignition sites occur during knocking combustion. The combustion mode we have monitored we believe differs from knocking combustion in terms of the size, gradients, and spatial distribution of the exothermic centers in the end-gas. This combustion concept was given a name PREMIER combustion.

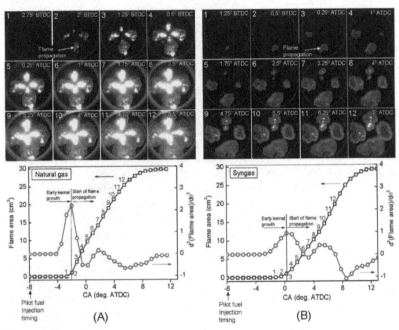

Figure 3. Dual-fuel combustion sequential images at P_{inj} = 40 MPa, P_{in} = 101 kPa, θ_{inj} = 8° BTDC, m_{df} = 2 mg/cycle, ϕ_t = 0.6. (A) natural gas, CH_4=88%, C_2H_6=6%, C_3H_8=4%, n-C_4H_{10}=2% (B) syngas, H_2=22.3%, CO=27.6%, CH_4=2.7%, CO_2=23.2%, N_2=24.2%

A conceptual outline of PREMIER combustion is presented in Figure 4. In the first stage of this combustion mode, the pilot diesel fuel is injected, evaporated, and auto-ignited prior to top dead center (TDC). The energy released by the diesel fuel auto-ignition initiates the gaseous flame development and outward propagation from the ignition centers toward the cylinder wall. Once the end-gas region is sufficiently heated and the temperature of the fuel mixture has reached the auto-ignition temperature of the gaseous fuel/air mixture after TDC, the second-stage combustion begins and is completed as the gas expands and cools, producing work. The second-stage heat release occurs over a chemical reaction timescale and is faster than heat release by turbulent flame propagation. Thus, the combustion transition from the first stage to the second stage takes place when the overall heat release rate changes from the slower first-stage flame rate to the faster second-stage rate, and that transition is here measured as the point where the second derivative of the heat release rate is maximized, as shown in Figure 4.

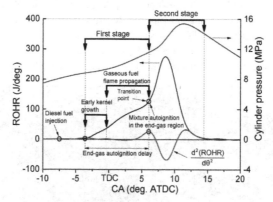

Figure 4. PREMIER combustion concept

PREMIER combustion in a dual-fuel engine is comparable to combining SI and CI combustion, which is being investigated by several researchers [24-26]. One disadvantage of these combustion strategies is that they are difficult to control under lean mixture conditions. The spark discharge is very short, and under light load and lean mixture conditions, the flame is too weak to propagate strongly and may be extinguished. Therefore, the combustion chamber must be specially designed to facilitate a stratified mixture charge around the spark plug electrodes. In a dual-fuel engine, on the other hand, combustion of the injected diesel fuel proceeds concurrently with that of the gaseous fuel mixture. This slow combustion of the diesel fuel helps to maintain the natural gas flame propagation and prevents the misfires that may occur under lean mixture conditions. The lean limit for the gaseous fuel/air mixture is of practical importance, as lean operation can result in both higher efficiency and reduced emissions. The major benefit of lean operation is the accompanying reduction in combustion temperature, which leads directly to a significant reduction in NOx emissions. The lean limit is the point where misfire becomes noticeable, and it is usually described in terms of the limiting equivalence ratio ϕ_{lim} that supports complete combustion of the mixture. For example, with the natural gas, if we operate slightly above the limiting equivalence ratio (for methane, $\phi_{lim} \approx 0.5$ [27]), the mixture reactivity becomes very sensitive to even very small variations in the air–fuel ratio. This high sensitivity is due to the presence of n-butane in the natural gas. It is known that small changes in the volume fraction of n-butane strongly affect the ignition properties of natural gas [28-31]. During hydrocarbon fuel oxidation, an H-atom is more easily abstracted from an n-butane molecule (with two secondary carbon atoms) than from other hydrocarbons such as methane, ethane, or propane [32]. As the equivalence ratio increases, the n-butane mass fraction in the natural gas/air mixture increases proportionally. During a fuel oxidation reaction, the in-cylinder gas temperature rises, and more of the radicals that initiate methane oxidation are created by increasing the ratio of n-butane. Similar results were documented by other researchers who investigated the auto-ignition and combustion of the natural gas in an HCCI engine [33]. They found that very small increases in the equivalence ratio of the methane/n-butane/air mixture produced significant changes in the profiles of the in-cylinder pressure traces.

If not extinguished during the early combustion stage, the pilot diesel flame may continue to a later stage, and pilot flame energy contributes to the stability of the combustion process [34]. The remaining unburned in-cylinder mixture from the first stage, located beyond the boundary of the flame front, is then subjected to a combination of heat and pressure for certain duration. As the flame front propagates away from the primary ignition points, end-gas compression raises the end-gas temperature and pressure. When the mixture is preheated throughout the combustion chamber volume and the end-gas mixture reaches the auto-ignition point, simultaneous auto-ignition occurs in several limited locations (known as exothermic centers), with a sharp increase in heat release. This part of the combustion appears as a rapid energy release in the second stage of the heat release curve shown in Figure 4.

A prime requirement for maintaining PREMIER combustion mode in a dual-fuel engine is that the mixture must not auto-ignite spontaneously during or following the rapid release of pilot energy. Failure to meet this requirement can lead to the onset of knock, which manifests itself in excessively sharp pressure increases and overheating of the walls, resulting in significant loss of efficiency with increased cyclic variations. When much smaller pilots are used, the energy release during the initial stages of ignition and the resulting turbulent flame propagation can (under certain conditions) lead to auto-ignition of the charge well away from the initial ignition centers, in the end-gas regions ahead of the propagating flames. This can occur in a manner that resembles the occurrence of knock in spark-ignition engines, but with controlled heat release. For the sake of convenience, the total energy release rate during PREMIER combustion can be divided into three sequential components. The first of these is due to the ignition of the pilot fuel. The second is due to the combustion of the gaseous fuel in the immediate vicinity of the ignition centers of the pilot, with consequent flame propagation. The third is due to auto-ignition in the end-gas region.

4. PREMIER combustion detection (Natural gas)

4.1. Cyclic variations and FFT of in-cylinder pressure

Cycle-to-cycle variations of the in-cylinder pressure during conventional, PREMIER, and knocking combustion with natural gas are identified by over-plotting 80 cycles of a measured pressure trace. From Figure 5, we observe that cycle-to-cycle pressure variations were present in all cases considered, and they varied according to the injection timing θ_{inj}. It should be noted that under normal combustion conditions, the magnitude of the peak pressure (which is directly related to the power output) depends on θ_{inj}, and P_{max} is higher for advanced θ_{inj}. Unfortunately, the advantage of this peak pressure increase is offset by the disadvantages of increased fluctuation and the occurrence of knock.

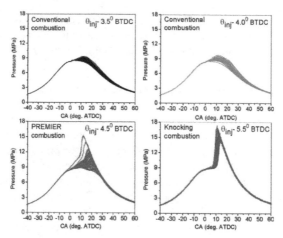

Figure 5. Cyclic maximum pressure versus its angle (Pmax, θPmax) for different injection timings. P_{inj}=80 MPa, P_{in}=200 kPa, D_{hole}= 0.1 mm, N_{hole}=3, m_{df}=2 mg/cycle, ϕ_t=0.6

Thus, when slow combustion dominates, the P_{max} fluctuations are small, whereas during fast combustion, the P_{max} fluctuations are larger. With advanced fuel injection timing, the combustion was too fast, and the mixture auto-ignited spontaneously during or following the rapid pilot energy release. This led to the onset of knock. However, with a slight retardation of the injection timing, the energy release during the initial stages of ignition and the resulting turbulent flame propagation may induce auto-ignition of the charge in the end-gas regions ahead of the propagating flame, followed by PREMIER combustion. The stability of PREMIER combustion was confirmed by running an engine mentioned in Figure 1 (A) continuously for 30 minutes. As the engine operation reached a steady condition, P_{max} stabilized within a definite range between conventional and knocking combustion modes. It should be noticed that the present achieved stability of PREMIER combustion allows using this mode in stationary engines, such as engines used for power generation. In order to utilize PREMIER combustion in vehicles, further research is required to find optimum ways to precisely control the conditions inside the cylinder at various loads.

The transition from PREMIER combustion to knocking combustion was evaluated by fast Fourier transform (FFT) analysis of the in-cylinder pressure and the knock-sensor signal. The details of FFT analysis are given in [35]. Unlike during knocking in traditional spark-ignition engines, high frequency oscillations of in-cylinder pressure and knock-sensor signals were not observed during the PREMIER combustion mode.

The in-cylinder pressure and knock-sensor signals were filtered using an FFT band-pass filter with 4-20 kHz cutoff frequencies. The filtered pressure data was used to define the knock intensity. Figure 6 shows that during PREMIER combustion mode, the measured cylinder pressure curve was smooth, and the maximum rate of pressure rise (which was only 0.36 MPa/CA deg.) occurred at 17° ATDC under the conditions P_{int}=200 kPa, m_{df}=2 mg/cycle, P_{inj}=80 MPa and θ_{inj}=4.5° BTDC. Pressure oscillations did not occur, and the pressure sensor was unable to detect a sawtooth pattern on the measured pressure trace. For the same conditions mentioned above, but at $\theta_{inj}=5°$ BTDC, the transition from PREMIER to knocking combustion occurred at the maximum rate of pressure increase ($dP/d\theta=0.55$ MPa/CA deg.), with a knock intensity of $K_{INT}=0.3$ MPa. At $\theta_{inj}=5.5°$ BTDC, strong knocking combustion was detected with $dP/d\theta=1.87$ MPa/CA deg. and $K_{INT}=1.45$ MPa. As the figure indicates, oscillations of the in-cylinder pressure and the knock sensor signal were not detected during PREMIER combustion mode. Weak oscillations were detected during the transition from PREMIER to knocking combustion, and stronger oscillations were detected at 6.52 kHz and 10.1 kHz during knocking combustion. The peak at the transition to knocking combustion was correspondingly lower than the peak that occurred during heavy knocking combustion.

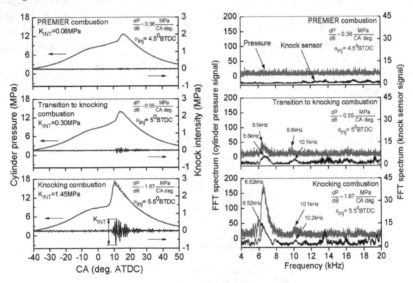

Figure 6. FFT analysis of in-cylinder pressure. $P_{inj}=80$ MPa, $P_{in}=200$ kPa, $D_{hole}=0.1$ mm, $N_{hole}=3$, $m_{df}=2$ mg/cycle, $\phi_t=0.6$

4.2. Spectroscopy analysis

The configuration of our optical engine made it difficult to directly visualize auto-ignition in the end-gas region with an optical engine setup shown in Figure 1 (B). Since auto-ignition in the end-gas region usually occurs closer to the cylinder wall, the auto-ignition region was hidden from the camera view by the top of the elongated piston, where a sapphire window was installed. Thus, an optical sensor was inserted in the region based on the most probable occurrence of auto-ignition in the end-gas region, as shown in Figure 7. The small effect of measurement location on autoignition and flame development was confirmed based on several preliminary experiments. Besides, this measurement location was selected to keep the sensor away from the highly luminescent soot radiation. If the sensor is placed too close

to the luminous emissions from diesel combustion, these emissions can supersede the OH*
radical emissions, which are expected to occur during PREMIER combustion. Chemical
luminescence emissions from the propagating flame and end-gas region auto-ignition were
measured using a spectrometer equipped with intensified charge-coupled device (ICCD).
The regions of spectroscopy measurements for conventional, PREMIER and knocking
combustion, along the crank angle degrees, are shown in Figure 8. In-cylinder pressure and
the rate of pressure rise change as injection timing is advanced for conventional, θ_{inj}=4°
BTDC, PREMIER, θ_{inj}=8° BTDC and knocking, θ_{inj}=13° BTDC, combustion. Exposure time of
the spectrometer for all three combustion regimes was set to 3° CA. The highlighted region
on each graph shows the interval within which the measurements were taken.

Figure 9 shows the background light-subtracted ensemble-averaged spectra obtained
between 200 and 800 nm as the propagating flame arrived in the vicinity of the optical
sensor and auto-ignition occurred in the end-gas region. The background light was
subtracted by the signal processing using a percentile filter. The waveforms of the emission
spectroscopy were obtained for knocking cycles in the range of 0°-3°, 3°-6°, and 6°-9° ATDC,
for PREMIER cycles in the range of 6°-9°, 9°-12°, and 12°-15° ATDC, and for conventional
cycles in the range of 12°-15°, 15°-18°, and 18°-21° ATDC.

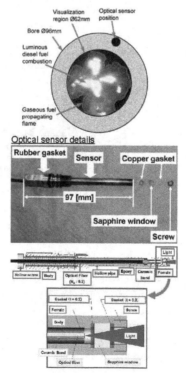

Figure 7. Optical sensor location and design

High OH* radical emission intensities were evident at wavelength 310 nm and very weak intensities at wavelength 286 nm when the flame front reached the optical sensor location region. OH* radical emission intensities at this wavelengths were stronger for knocking combustion cycles than for PREMIER combustion cycles, and for conventional combustion cycles these emission intensities were not seen. A similar trend was observed by Itoh et al. [36] and Hashimoto et al. [37]. These authors reported that under non-knocking operation, OH* radicals exhibit comparatively weak emission intensities. However, the OH* radical emission intensity gradually increased for autoignition and knocking cycles in comparison with a conventional cycle.

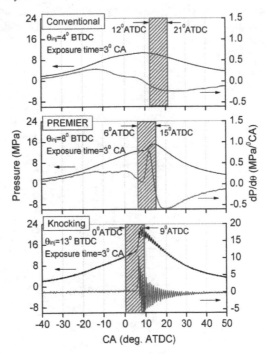

Figure 8. In-cylinder pressure and the rate of pressure rise. P_{inj}=40 MPa, P_{in}=200 kPa, D_{hole}=0.1mm, N_{hole}=3, m_{df}=2 mg/cycle, ϕ_t=0.6

Figure 9 shows that for the knocking cycle, the OH* peak at 310 nm increased sharply at 6° ATDC and then decreased at 9° ATDC. 6° ATDC corresponds to the timing of the maximum rate of pressure rise, and 9° ATDC corresponds to the timing when the rate of pressure rise is gradually decreased. Before natural gas propagating flame (1st stage combustion) is stabilized, the rapid autoignition occurs in multiple spots. The end-gas autoignition kernel growth rate is much faster than that of flame propagation.

The drastic increase in OH* emissions is related to the sharp pressure and temperature increases in the end-gas region. It has been reported that the concentration of OH* radicals shows a strong temperature dependence in the thermal ignition region [38]. The thermal ignition is the hydrogen-oxygen system reactions, and the CO conversion into CO_2 with the assistance of OH* takes place simultaneously with the hydrogen-oxygen system reactions. It has been observed that the OH* emission intensity at auto-ignition shows the same tendency as the correlation between the occurrence of auto-ignition and the knocking intensity [39]. On the other hand, the OH* peaks of PREMIER combustion at 12° and 15° ATDC have the same magnitude. 12° ATDC corresponds to the timing of the maximum rate of pressure rise, and 15° ATDC corresponds to the timing after the peak of the rate of pressure rise. The reason for the same magnitude in OH* is that the maximum rate of pressure rise is slower than that of during knocking case. The end-gas autoignition kernel growth rate is comparable with that of flame propagation. The mixture in the end-gas region reacts steadily with the steady OH*emission.

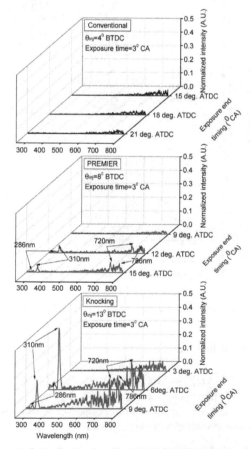

Figure 9. Detailed spectra in the end-gas region. P_{inj}=40 MPa, P_{in}=200 kPa, D_{hole}=0.1 mm, N_{hole}=3, m_{df}=2 mg/cycle, ϕ_t=0.6

The emissions at the wavelengths above 700 nm are due to diesel luminous flame. The previous shock tube measurements showed that continuum emission at wavelengths above 650 nm was due to either young soot particles or large hydrocarbon molecules [40]. Zhao and Ladommatos [41] showed that the maximum emissive power of soot particles, estimated as a blackbody, occurs in the range of about 680–1100 nm. Vattulainen *et al.* [42] confirmed that the light emitted by a diesel flame is dominated by soot incandescence, and the emission range was determined to be about 650–800 nm. The emissions at the wavelengths above 700 nm shown in Figure 9 correspond to 1-5% of the total wavelength emission band from a blackbody at the estimated average peak burned-gas temperature of 2310 K under knocking conditions, which correspond to 1.823×10^5-4.935×10^5 W/m^2·μm of spectral emissive power, with the maximum spectral emissive power at that temperature that equals to 8.459×105 W/m^2·μm [43].

5. PREMIER combustion characteristics

5.1. Natural gas combustion characteristics

To maintain PREMIER combustion in a dual-fuel natural gas engine, the effects of several operating parameters must be identified. Our experimental results show that the major parameters that may significantly influence the energy release pattern during dual-fuel PREMIER combustion are pilot diesel fuel injection timing and the EGR rate, which can affect the total equivalence ratio based on oxygen content. Other parameters such as pilot fuel injection pressure, injected pilot fuel amount, nozzle hole diameter, and hole number have minor effects on PREMIER combustion.

5.1.1. Effect of injection timing

As shown in Figure 10, PREMIER combustion can be maintained within a wide range of pilot fuel injection pressures. However, it can be maintained within only a very narrow range of fuel injection timings. At a fixed total equivalence ratio, advancing the injection timing resulted in the earlier occurrence of combustion during the cycle, increasing the peak cylinder pressure during first-stage combustion. With the burned gas of first-stage combustion, the in-cylinder pressure and temperature continued to rise after TDC, as shown in Figure 10. Although the piston began to move downward after TDC, and the volume thus expanded, the heat release from first-stage combustion induced local temperature and pressure increases during second-stage combustion. Higher peak cylinder pressures resulted in higher peak charge temperatures. Retarding the injection timing decreased the peak cylinder pressure during first-stage combustion, as more of the fuel burned after TDC.

Advancing the injection timing resulted in better diesel fuel evaporation and mixing with the in-cylinder gas. Therefore, diesel fuel auto-ignition occurred more quickly and with more complete diesel fuel combustion and natural gas flame propagation during the first stage, resulting in rapid combustion and high heat release rate during the second stage due to the rapid heating of the end-gas region mixture.

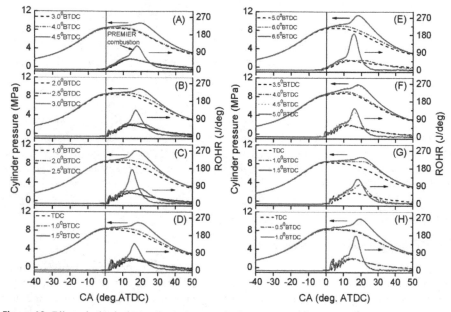

Figure 10. Effect of pilot fuel injection timing on cylinder pressure and the rate of heat release. Experimental conditions are given in Table 3.

	Injection pressure P_{inj} (MPa)	Intake pressure P_{in} (kPa)	Nozzle hole diameter D_{hole} (mm)	Nozzle hole number N_{hole}	Injected pilot fuel amount m_{df} (mg/cycle)
A	40	200	0.1	3	3
B	80	200	0.1	3	3
C	120	200	0.1	3	3
D	150	200	0.1	3	3
E	40	200	0.1	3	2
F	80	200	0.1	3	2
G	150	200	0.08	3	3
H	150	200	0.1	4	3

Table 3. Experimental conditions for Figure 10

5.1.2. Mass fraction burned in the second stage of ROHR

Figure 11 shows a relation between the rate of maximum pressure rise and the second stage autoignition delay. The second stage autoignition delay time was estimated as a time between two peaks of the second derivative of the ROHR, as shown in Figure 4. Figure 11 also shows that for the range of experimental conditions the mass fraction burned during the second stage of the ROHR is remained nearly the same, although IMEP increases. This increase in IMEP is due to faster and more intense combustion with the shorter duration in both first and second stage of the ROHR. Reduced heat loss during shorter combustion duration ensures higher in-cylinder pressure and temperature, and therefore, higher IMEP. This trend suggests that the second stage combustion is influenced by the first stage. Although the mass fraction burned in the second stage is the same, the combination of several operational parameters such as pilot fuel injection pressure, injection timing, the amount of injected pilot fuel, gaseous fuel equivalence ratio and nozzle characteristics may affect, to a certain degree, the progress of PREMIER combustion.

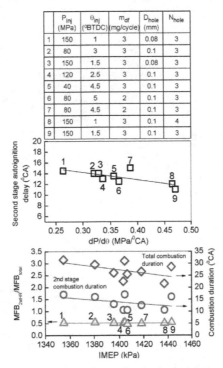

Figure 11. Second stage autoignition delay and mass fraction burned during the second stage of combustion of natural gas

5.1.3. Effect of EGR

PREMIER combustion becomes clearly recognizable if the EGR rate remains below a certain level. When the EGR rate surpasses this level, the unburned mixture temperature decreases, retarding the combustion of the natural gas and affecting the reactivity of the mixture to auto-ignite in the end-gas region. As Figure 12 shows, at 200 kPa of intake pressure and moderate EGR rates, the first-stage combustion rate increased, and second-stage heat release was able to occur. A similar trend was also observed at 101 kPa of intake pressure, although not shown here. These results suggest that the use of EGR may not be advantageous for achieving PREMIER combustion. However, it should be noted that for engine operation close to the knock-limit conditions, the high combustion rate of natural gas may be markedly decreased by using a certain limited EGR rate, and maintaining PREMIER combustion mode as the knocking effect is suppressed.

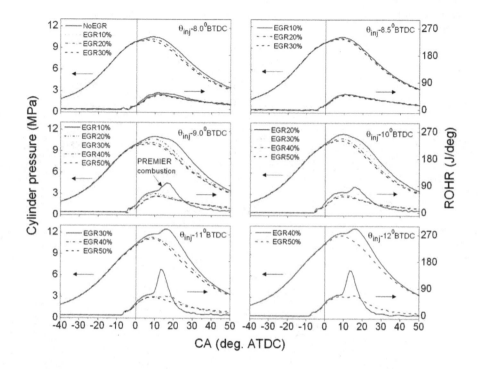

Figure 12. Effect of EGR on cylinder pressure and the rate of heat release. P_{inj}=40 MPa, P_{in}=200 kPa, D_{hole}=0.1 mm, N_{hole}=3, m_{df}=2 mg/cycle

5.2. Natural gas exhaust emission characteristics

Figure 13 shows the engine performance characteristics and exhaust gas emissions. The encircled data on (A) and (B) of the figure correspond to PREMIER combustion. During PREMIER combustion, a considerable increase in the indicated mean effective pressure and thermal efficiency was observed. This was due to the sharp increase of the second heat-release peak within a shorter crank-angle time, as seen in Figures 10 and 12. Moreover, this implies that the total combustion time for PREMIER combustion mode was shorter than the total time required for conventional combustion. Although HC and CO emissions were greatly decreased, NOx emissions increased considerably compared with conventional combustion. This increase in NOx emissions is expected, as in-cylinder temperature increase hastens the oxidation reactions of in-cylinder nitrogen and oxygen.

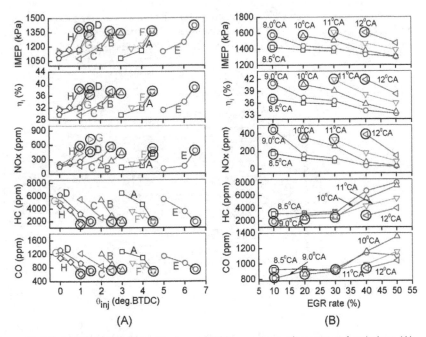

Figure 13. Effect of pilot fuel injection timing and EGR on engine performance and emissions. (A) conditions correspond to those of Figure 10, (B) conditions correspond to those of Figure 12.

5.3. Syngas combustion characteristics

Figure 14 indicates the relationship between the rate of maximum pressure rise and the second stage autoignition delay. The second stage autoignition delay time was estimated as time between two peaks of the second derivative of the ROHR as shown in Figure 4. It was found that for all types of gases investigated in this chapter the autoignition delay of the second stage decreases with the increase of the rate of maximum pressure rise.

Figure 15 shows the in-cylinder pressure and ROHR for Type 1 (A) and Type 2 (B) of syngas at different equivalence ratios and various injection timings. The results show that the maximum pressure and heat release rate reached higher values for the gas with higher H_2 content. As the injection timing was gradually advanced, PREMIER combustion with second-stage heat release occurred. PREMIER combustion was observed at various equivalence ratios for certain injection timings. For instance, for Type 1 gas, two-stage heat release appeared at $\phi = 0.4$ starting from 23° BTDC, at $\phi = 0.52$ from 15° BTDC, at $\phi = 0.68$ from 9° BTDC, and at $\phi = 0.85$ from 7° BTDC. The same trend was observed for Type 2 gas, but the maximum heat release rate was higher than that of Type 1, due to the higher H_2 content. The maximum cylinder pressure for Type 2 decreased at an equivalence ratio of 0.83, since the injection timing in that case needed to be retarded to around TDC (and even to the expansion stroke) to avoid knock. At the same time, for Type 1 gas at an equivalence ratio of

0.85 and injection timings before TDC, PREMIER combustion was clearly observed without any knock. These results seem to suggest that the increased H_2 content of Type 2 gas affected the ignitability and corresponding progress of combustion that leads to engine knock.

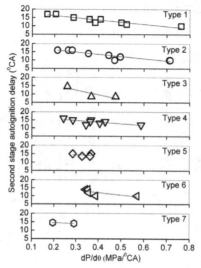

Figure 14. Second stage autoignition delay. P_{inj}=80 MPa, P_{in}=200 kPa, m_{df}=3 mg/cycle

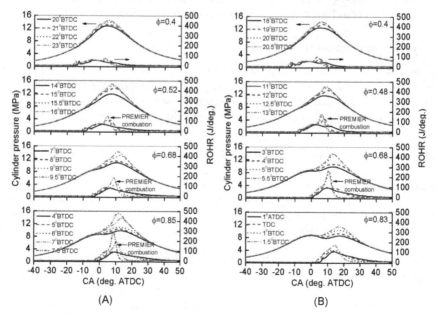

Figure 15. Comparison of in-cylinder pressure and ROHR. Syngas, (A) Type 1, (B) Type 2

5.3.1. Effect of H_2 content on PREMIER combustion

An increase in hydrogen content of syngas results in an increase of ignitability and a corresponding reduction in ignition delay of the first stage. To investigate the effect of mass fraction burned in the second stage and the effect of hydrogen content on the rate of maximum pressure rise, the ratio MFB_{2stHR}/MFB_{total} was evaluated. MFB_{2stHR} is the integral of the mass fraction burned during the second stage, computed from the transition point where the peak of $d^2(ROHR)/d\theta^2$ is maximized to the 80% MFB. MFB_{total} is the integral of the total mass fraction burned, computed from the first peak of $d^2(ROHR)/d\theta^2$ to the 80% MFB. Figure 16 shows that as the mass fraction burned increases the rate of maximum pressure rise also increases. The same trend was monitored at various equivalence ratios.

Figure 17 shows the effect of H_2 content on the mean combustion temperature, IMEP, indicated thermal efficiency and NOx for conventional and PREMIER combustion at the same input energy Q_{in}=2300 J/cycle and injection timing at the minimum advance for the best torque (MBT). As this figure shows, the increase in H_2 amount affects the engine combustion characteristics. For PREMIER combustion, the mean combustion temperature and consequently the NOx significantly increase when compared with those of conventional combustion. IMEP and indicated thermal efficiency increase about 10%. Therefore, in dual-fuel combustion of low-energy density syngas, PREMIER combustion is an important combustion mode that tends to increase the engine efficiency.

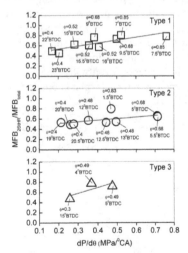

Figure 16. Fuel mass fraction burned with the change of H_2 content

5.3.2. Effect of CO_2 content on PREMIER combustion

An increase in CO_2 content in syngas results in a dilution of the mixture with the corresponding reduction in the rate of fuel oxidation reactions and consequent combustion. To investigate the effect of mass fraction burned in the second stage and the effect of CO_2

content on the rate of maximum pressure rise, the same procedure was applied as explained in the previous section.

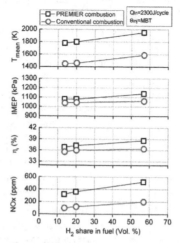

Figure 17. Effect of H₂ content on engine performance characteristics at θinj=MBT

Figure 18 shows that for Type 1 and Type 4 as the mass fraction burned during the second stage increases the rate of maximum pressure rise also increases. However, for Type 5 with 34% of CO₂ content in syngas, the rate of maximum pressure rise decreases with the increase of the mass of fuel burned in the second stage. This can be explained by the fact that although the total mass of syngas burned during the second stage increases, the CO₂ fraction in the gas also proportionally increases. Eventually, the certain threshold can be reached when the effect of CO₂ mass fraction in the gas on combustion overweighs the effect of H₂.

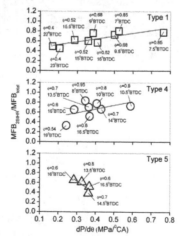

Figure 18. Fuel mass fraction burned with the change of CO₂ content

Figure 19 shows the effect of CO_2 content on the mean combustion temperature, IMEP, indicated thermal efficiency and NOx for conventional and PREMIER combustion at the same input energy Q_{in}=2300 J/cycle and injection timing at MBT. In this figure the trend mentioned earlier in Figure 18 is clearly observed.

The mean combustion temperature, IMEP, the indicated thermal efficiency and the NOx shows the increase when CO_2 content in the gas increases from 16.8% to 23%. However, as CO_2 concentration reaches 34%, above mentioned combustion characteristics decrease. This trend was observed for both conventional and PREMIER combustion. Therefore, in order to achieve high combustion efficiencies in dual-fuel engines fuelled with low-energy density syngas, CO_2 fraction in syngas needs to be controlled.

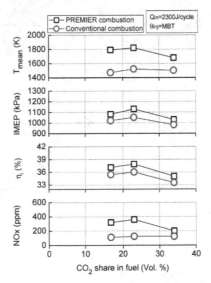

Figure 19. Effect of CO_2 content on engine performance characteristics at θ_{inj}=MBT

5.4. Syngas exhaust emission characteristics

Increasing hydrogen content in syngas has a substantial impact on engine performance and pollutants formation. Engine performance characteristics and concentrations of pollutants CO, HC and NOx are shown in Figures 20 and 21. It should be noticed that the smoke level was negligibly low. Figure 3 (B) shows highly luminous diesel fuel droplets burn exposed to the high temperature syngas flame. This concurrent micro-pilot diesel fuel and syngas combustion contributes to the faster oxidation and burn out of soot by the end of the combustion process.

Figures 20 and 21 show the experimental operation region for IMEP, CO, HC and NOx. The comparison is given for H_2 effect between Type 1 and Type 2, for CO_2 effect between Type 1 and Type 5, and for gas - type effect between Type 6 (Coke-oven gas) and Type 8 (Pure hydrogen). The dash-dotted line is the boundary separated the conventional combustion from the PREMIER combustion at corresponding equivalence ratios and injection timings.

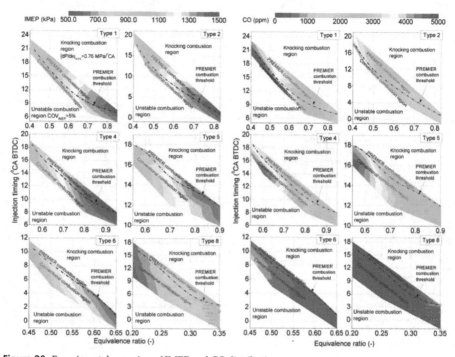

Figure 20. Experimental mapping of IMEP and CO distribution

Figure 20 shows that higher IMEP levels appear at higher equivalence ratios and advanced injection timings. PREMIER combustion mode may pass through different IMEP levels, depending on the equivalence ratio and injection timing. As H_2 fraction increases in the syngas (from Type 1 to Type 2), PREMIER combustion region expands at lower equivalence ratios and shrinks at higher equivalence ratios. On the contrary, as CO_2 fraction increases in the syngas, from Type 1 (16.8% CO_2) to Type 5 (34% CO_2), the same maximum level of IMEP, as for Type 1 and 2, can be achieved with only higher equivalence ratios. Type 1 and Type 5 show that as CO_2 content increases IMEP slightly decreases. With the increase of CO_2 fraction in the syngas the PREMIER combustion region threshold is shifted towards the boundary of operation domain. This implies that the operational region of PREMIER combustion mode is reduced.

Figure 20 (Type 1 and Type 2) and Figure 21 (Type 1 and Type 2) show that with the increase of H_2 concentration in the gas content, CO and HC emissions are reduced. The cause of these reductions is most likely the enhanced oxidation occurring because of improved combustion and higher concentrations of reactive radicals since the carbon content in the gas is the same for both types of fuel. The NOx emissions, as in Figure 21, are higher for Type 2 than those for Type 1 due to higher H₂ content. This is in part a direct result of hydrogen's higher flame temperature effect on NO formation chemistry. NOx emissions are consistently reduced by lowering equivalence ratio for both types. For Type 5, as CO₂ content in the gas increases to 34%, the NOx reduction trend is obvious.

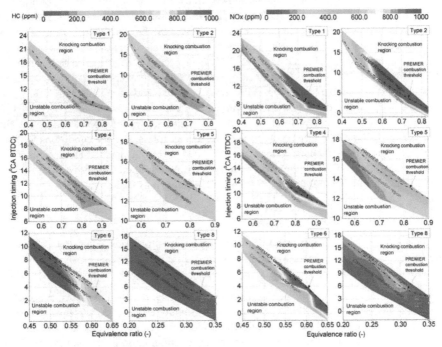

Figure 21. Experimental mapping of HC and NOx distribution

The opposite trend is observed for Type 6, as shown in Figure 20. This type of gas has very high content of H₂-56.8%, and very low content of CO₂-2.2%. For the limited range of equivalence ratios and injection timings, the PREMIER combustion region is very narrow, and at even higher equivalence ratios, the stable combustion will easily turn to knocking. These results show that in order to achieve PREMIER combustion with the best case scenario in terms of efficiency and emissions the equivalence ratio and injection timing should be maintained within certain range, depending on H₂ and CO₂ fractions in syngas. Figure 21 shows that for Type 6 the NOx emissions can be compared with those of Type 2. Even with the retarding of the injection timing closer to TDC, NOx emissions level is still comparably high. On the other hand, CO emissions are significantly reduced.

In addition, the performance and emissions of pilot-ignited dual-fuel engine operated on 100% H₂ as a primary fuel was investigated, shown as Type 8. The data at equivalence ratios of 0.2, 0.25 and 0.3 are related to the engine operation without dilution. Further increase in equivalence ratio above 0.3 caused knocking combustion. To further increase the energy supply from hydrogen the equivalence ratio of 0.35 was used with the 40%, 50% and 60% of N_2 dilution. Previous research suggests that NOx emissions in hydrogen-fueled engines can be reduced by using EGR or using the nitrogen dilution [44, 45]. These studies reported that NOx emissions were reduced by the reduction in peak combustion temperature due to the presence of diluent gas. Mathur *et al.* [45] showed that with larger concentrations of nitrogen as diluent, a greater amount of combustion heat can be absorbed, which results in reducing peak flame temperature and NOx. IMEP increases with the equivalence ratio. For the equivalence ratios of 0.2 and 0.25, the minimum advanced injection timing for the maximum IMEP was obtained at 17° BTDC and 12° BTDC, respectively. At these injection timings PREMIER combustion occurs. For the case with N_2 dilution, injection timing has greater effect on IMEP than the dilution rate. IMEP gradually increased as the injection timing was advanced along with the increased amount of diluent from 40% to 60%. Due to low equivalence ratios, very low level of NOx was detected without N_2 dilution. At the conditions with N_2 dilution the NOx showed further decrease. For the conditions without dilution, CO and HC emissions were significantly reduced to the level of about 6 ppm and 20 ppm, respectively. However, for the conditions with N_2 dilution CO varied from min.-15 ppm to max.-891 ppm and HC varied from min.-20 ppm to max.-125 ppm.

6. Conclusion

The new PREMIER combustion mode in a dual-fuel engine fuelled with natural gas, syngas and hydrogen was investigated via engine experiments. The following conclusions can be drawn from this research:

- PREMIER combustion combines two main stages of heat release, the first is gaseous fuel flame propagation and the second is end-gas mixture auto-ignition. The second stage can be mainly controlled by the pilot fuel injection timing, gaseous fuel equivalence ratio, and EGR rate. The delay time for mixture autoignition in the end-gas region is defined as the time from early kernel development to the transition point where slower combustion rate (flame propagation) is changed to faster combustion rate (autoignition). It was found that the rate of maximum pressure rise increases as the second stage ignition delay decreases. PREMIER combustion was observed for natural gas and all syngas types investigated in this paper. This type of combustion can enhance the engine performance and increase the efficiency.
- In both PREMIER and knocking combustion with natural gas, moderate and high intensities of OH* radicals were detected, respectively, at wavelengths of 286 and 310 nm when the flame front reached the optical sensor location region. In knocking combustion, the OH* radical emission intensities were stronger than those in PREMIER combustion, and in conventional combustion, these emission intensities could barely be detected. PREMIER combustion differs from knocking combustion in terms of size, gradients, and spatial distribution of exothermic centers in the end-gas region. High-frequency oscillation of the in-cylinder pressure did not occur during the PREMIER combustion mode with natural gas.

- An increase in the fuel mass fraction burned in the second stage of heat release affects the rate of maximum pressure rise. When hydrogen content in syngas is increased the same rate of maximum pressure rise can be achieved with lower amount of fuel mass fraction burned during the second stage, meaning that the increased amount of hydrogen in syngas induces an increase in the mean combustion temperature, IMEP and efficiency, but also a significant increase in NOx emissions. The results also show that when CO_2 content in the gas reaches 34%, the rate of maximum pressure rise, as well as, the mean combustion temperature, IMEP, thermal efficiency and NOx decrease despite the increase in fuel mass fraction burned during the second stage of heat release rate.
- For pure hydrogen at equivalence ratios of 0.2, 0.25 and 0.3 without dilution, very low CO and HC emissions were detected. The further increase in equivalence ratio above 0.3 led to knocking combustion. At the equivalence ratio of 0.35 with N_2 dilution, NOx level significantly decreased but CO level increased.

Author details

Ulugbek Azimov*
Department of Mechanical Engineering,Curtin University, Malaysia campus

Eiji Tomita and Nobuyuki Kawahara
Department of Mechanical Engineering, Okayama University, Japan

7. References

[1] Weaver CS. Natural gas vehicles – a review of the state of the art, SAE Paper, 892133.

[2] Nichols RJ. The challenges of change in the auto industry: Why alternative fuels? J Eng Gas Turb Power 1994;116:727-32.

[3] Lieuwen T, Yang V, Yetter R. Synthesis gas combustion: Fundamentals and applications. Taylor & Francis Group 2010.

[4] Shilling NZ, Lee DT. IGCC-Clean power generation alternative for solid fuels: GE Power Systems. PowerGenAsia 2003.

[5] Bade Shrestha SO, Karim GA. Hydrogen as an additive to methane for spark ignition engine applications. Int J Hydrogen Energy 1999;24:577-86.

[6] Stebar RF, Parks FB. Emission control with lean operation using hydrogen-supplimented fuel. SAE Paper 740187.

[7] Jingding L, Linsong G, Tianshen D. Formation and restraint of toxic emissions in hydrogen-gasoline mixture fueled engines. Int J Hydrogen Energy 1998;23:971-75.

[8] Pushp M, Mande S. development of 100% producer gas engine and field testing with pid governor mechanism for variable load operation. SAE Paper 2008-28-0035.

---

* Corresponding Author

[9] Yamasaki Y, Tomatsu G, Nagata Y, Kaneko S. Development of a small size gas engine system with biomass gas (combustion characteristics of the wood chip pyrolysis gas), SAE Paper 2007-01-3612.

[10] Ando Y, Yoshikawa K, Beck M, Endo H. Research and development of a low-BTU gas-driven engine for waste gasification and power generation. Energy 2005;30:2206-18.

[11] McTaggart-Cowan GP, Jones HL, Rogak SN, Bushe WK, Hill PG, Munshi SR. The effect of high-pressure injection on a compression-ignition, direct injection of natural gas engine. J Eng Gas Turb Power 2007;129: 579-88.

[12] Su W, Lin Z. A study on the determination of the amount of pilot injection and rich and lean boundaries of the pre-mixed CNG/Air mixture for a CNG/Diesel dual-fuel engine. SAE paper, 2003-01-0765.

[13] Tomita E, Fukatani N, Kawahara N, Maruyama K, Komoda T. Combustion characteristics and performance of supercharged pyrolysis gas engine with micro-pilot ignition. CIMAC congress 2007. Paper No. 178.

[14] Liu Z, Karim GA. Simulation of combustion processes in gas-fuelled diesel engine. J Power Energy 1997;211:159-69.

[15] Ma F, Wang Y, Ding S, Jiang L. Twenty percent hydrogen-enriched natural gas transient performance research. Int J Hydrogen Energy 2009;34:6423-31.

[16] Sung CJ, Law CK. Fundamental combustion properties of H2/CO mixtures: Ignition and flame propagation at elevated temperatures. Combust Sci.Tech. 2008;180:1097-1116.

[17] Heywood JB. Internal combustion engine fundamentals. New-York: McGraw-Hill 1988.

[18] Nakagawa Y, Takagi Y, Itoh T, and Iijima T. Laser shadowgraphic analysis of knocking in SI engine. SAE Paper, 845001.

[19] Pan J, Sheppard CGW, Tindall A, Berzins M, Pennington SV and Ware JM. End-gas inhomogeneity, autoignition and knock. SAE Paper, 982616.

[20] Kawahara N, Tomita E, and Sakata Y. Auto-ignited kernels during knocking combustion in a spark-ignition engine. Proc. Combust. Inst. 2007;31:2999-3006.

[21] Bauerle B, Hoffman F, Behrendt F, and Warnatz J. Detection of hot spots in the end gas of an internal combustion engine using two-dimensional LIF of formaldehyde. In: 25th Symposium (Int.) on Combustion 1994;25:135-141.

[22] Stiebels B, Schreiber M, and Sadat Sakak A. Development of a new measurement technique for the investigation of end-gas autoignition and engine knock. SAE Paper, 960827.

[23] Pan J, and Sheppard CGW. A theoretical and experimental study of the modes of end gas autoignition leading to knock in SI engines. SAE Paper, 942060.

[24] Li J, Zhao H, Ladommatos N. Research and development of controlled auto-ignition (CAI) combustion in a 4-stroke multi cylinder gasoline engine. SAE Paper, 2001-01-3608.

[25] Santoso H, Matthews J, Cheng WK. Managing SI/HCCI dual-mode engine operation. SAE Paper, 2005-01-0162.

[26] Persson H, Hultqvist A, Johansson B, Remon A. Investigation of the early flame development in spark assisted HCCI combustion using high-speed chemiluminescence imaging. SAE Paper, 2007-01-0212.

[27] Ma F, Wang J, Wang Yu, Wang Y, Li Y, Liu H, Ding S. Influence of different volume percent hydrogen/natural gas mixtures on idle performance of a CNG engine. Energy & Fuels 2008;22:1880-87.

[28] Higgin RMR, Williams A. A shock-tube investigation of the ignition of lean methane and n-butane mixtures with oxygen. In: Symposium (Int.) on Combustion 1969;12:579-90.

[29] Zellner R, Niemitz KJ, Warnatz J, Gardiner Jr WC, Eubank CS, Simmie JM. Hydrocarbon induced acceleration of methane-air ignition. Prog Astronaut Aeronaut 1981;88:252-72.

[30] Eubank CS, Rabinovitz MJ, Gardiner Jr WC, Zellner RE. Shock-initiated ignition of natural gas-air mixtures. In: 18th Symposium (Int.) on Combustion 1981;18:1767-74.

[31] Crossley RW, Dorko EA, Scheller K, Burcat A. The effect of higher alkanes on the ignition of methane-oxygen-argon mixtures in shock waves. Comb. Flame 1972;19:373-78.

[32] Westbrook CK, Pitz WJ, Leppard WR. The autoignition chemistry of paraffinic fuels and pro-knock and anti-knock additives: a detailed chemical kinetic study. SAE Paper, 912314.

[33] Jun D, Ishii K, Iida N. Autoignition and combustion of natural gas in a 4-stroke HCCI engine. JSME International J. 2003;46:60-67.

[34] Saito H, Sakurai T, Sakonji T, Hirashima T, Kanno K. Study on lean-burn gas engine using pilot oil as the ignition source. SAE Paper, 2001-01-0143.

[35] Azimov U, Tomita E, Kawahara N, Harada Y. PREMIER (Premixed Mixture Ignition in the End-gas Region) combustion in a natural gas dual-fuel engine: Operating range and exhaust emissions. Int J Engine Research 2011;12:484-497.

[36] Itoh T, Nakada T, and Takagi Y. Emission characteristics of OH and C2 radicals under engine knocking. JSME International J. 1995;38:230-237.

[37] Hashimoto S, Amino Y, Yoshida K, Shoji H and Saima A. Analysis of OH radical emission intensity during autoignition in a 2-stroke SI engine. In: Proceedings of the 4th COMODIA 1998, 405-410.

[38] Gaydon AG. The spectroscopy of flames, 2nd edition, 1974 (Chapman and Hall Ltd, London).

[39] Itoh T, Takagi T, and Iijima T. Characteristics of mixture fraction burned with autoignition and knocking intensity in a spark ignition engine. Trans. Japan Soc. Mech. Eng. 1985;52:3068.

[40] Coats CM, and Williams A. Investigation of the ignition and combustion of n-Heptane-oxygen mixtures. Proc. Combust. Institute 1979;17:611-621.

[41] Zhao H, and Ladommatos N. Optical diagnostics for soot and temperature measurement in diesel engines. Prog. Energy Combust. Sci. 1998;24:221-255.

116

[42] Vattulainen J. Experimental determination of spontaneous diesel flame emission spectra in a large diesel engine operated with different diesel fuel qualities. SAE Paper, 981380.

[43] Incropera FP and DeWitt DP. Fundamentals of heat and mass transfer, 4th edition, 1996 (John Wiley & Sons).

[44] Bose PK, Maji D. An experimental investigation on engine performance and emissions of a single cylinder diesel engine using hydrogen as induced fuel and diesel as injected fuel with exhaust gas recirculation. Int J Hydrogen Energy 2009;34:4847-54.

[45] Mathur HB, Das LM, Patro TN. Hydrogen-fuelled diesel engine: Performance improvement through charge dilution techniques. Int J Hydrogen Energy 1993;18:421-31.

Chapter 5

Structured Catalysts for Soot Combustion for Diesel Engines

E.D. Banús, M.A. Ulla, E.E. Miró and V.G. Milt

Additional information is available at the end of the chapter

1. Introduction

During the last decade, diesel engines have increased in popularity compared to gasoline engines because of their superior fuel economy, reliability and durability, simultaneously associated with a favorable fuel tax situation in several countries. At present, around 60% of all new European cars are equipped with diesel engines [1,2]. However, the emission from this kind of engines includes important contaminants, nitrogen oxides (NO_x) and particulate matter (PM) being the ones of major concern. PM consists mostly of carbonaceous soot and a soluble organic fraction (SOF) of hydrocarbons condensed on the soot. Both PM and nitric oxides are known harmful compounds that affect human health and produce acid rain and photochemical smog, causing severe environmental problems.

The heterogeneity in the composition of the diesel fuel and air mixture in the combustion chamber of a diesel engine produces a great number of problems related to contaminant emissions. At the moment of the fuel injection, three regions can be distinguished in the cylinder: i) A region close to the injector, where the concentration of diesel fuel is relatively high; ii) The central region of the cylinder, where conditions are close to the stoichiometric ones; and iii) The most separated region from the injector, where the fuel concentration is low. If the generated turbulence is not enough in the first region, in the proximities of fuel drops there would not be sufficient oxygen to complete the fuel combustion. This leads to the formation of particulate matter (PM), which constitutes the typical black smoke observed from diesel vehicles during high acceleration or when they are cold. In contrast, in the third region, the oxygen excess and the very high temperature cause the appearance of important NO_x quantities. In the intermediate region, the diesel fuel/air ratio is close to the stoichiometric one so that the combustion is produced under near ideal conditions.

As previously said, PM is produced mainly close to the injector, in the areas where the fuel/air ratio is significantly high. Its formation involves four fundamental steps (Figure 1) [3].

1. *Pyrolysis*: A high concentration of fuel at high temperatures provokes that the fuel organic compounds alter their molecular structure without significant oxidation, although some oxygen may be present in the structure of the formed species. These molecules, such as C_2H_2, benzene and polycyclic aromatic hydrocarbons (PAHs) are called PM precursors.

2. *Nucleation*: Germs or nuclei of the PM primary particles are formed by arrangement of those precursors.

3. *Surface growing*: Primary particles are formed by a process of adding mass to the surface of a nucleated soot particle.

4. *Coalescence and agglomeration*: The combination of primary particles produces agglomerates, which are bigger particles than primary ones. During this process, from a greater number of small particles a smaller number of bigger particles are formed.

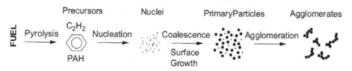

Figure 1. Steps involved in the formation of particulate matter (PM).

The PM composition depends on both the fuel quality and the heterogeneity of the combustion process occurring in the diesel engine. The PM is mostly composed of carbon, metal compound ashes and heavy organic compounds (soot), having adsorbed unburned hydrocarbons and condensed sulfates (SOF) on its surface.

The nitrogen oxides of the polluting emission gases are formed fundamentally at high temperatures in the low fuel/air ratio regions of the combustion chamber. These oxides are mainly nitric oxide (NO) and nitrogen dioxide (NO_2), and are collectively called NO_x.

The NO_x generation during combustion processes can take place through three different reaction paths, which have unique characteristics [4]:

1. *Thermal NO_x*: The reaction between atmospheric nitrogen and oxygen at high temperatures produces NO_x. The typical mechanism of this reaction was established by Zeldovich in 1946. The proposed mechanism involves a chain reaction of O* and N* activated atoms, NO being the final product. The conversion of NO to nitrogen dioxide (NO_2) occurs at low temperatures when exhaust gases are vented to the atmosphere.

2. *Fuel NO_x*: Fuel-bound nitrogen (FBN) is the source of NO_x emissions from combustion of nitrogen-bearing fuels such as heavy oils. FBN is converted to fixed nitrogen compounds such as HCN and NH_3 in the reducing region of the engine (high fuel/air ratio zone). And then, the latter compounds are oxidized to NO_x in the low fuel/air ratio region.

3. *Prompt NO_x*: In the fuel-rich region, atmospheric nitrogen can react with different hydrocarbon fragments (C, CH, CH_2), producing nitrogen containing intermediate species (NH, HCN, H_2CN, CN). Then, these species react with O_2 in the low fuel/air

ratio region. The prompt mechanism is responsible for only a small fraction of the total NO_x.

Legislations related to the control of pollutants emission are becoming more severe; therefore, industrial R&D departments and research groups face serious challenges to meet the specific emission requirements of future regulations. These days, most car manufacturers consider that a combination of catalysts and soot traps is the best approach to find a sustainable control of both NO_x and soot emissions [2].

The combination of a filter with oxidation catalysts is the most widely studied after treatment process to eliminate soot particles. For NO_x abatement, adsorbers (traps) and SCR with ammonia or hydrocarbons constitute technologies both for partial lean-burn gasoline engines and for diesel ones. The adsorber (usually containing Ba or K compounds) chemically binds nitrogen oxide during lean engine operation. After the adsorber capacity is saturated, the system is regenerated and released NO_x is reduced during a period of rich operation. In the case of diesel engines, since a rich operation is not feasible, periodic fuel injections are necessary [1].

2. Emission control

For diesel fuel, low volatility and excellent ignitability make it difficult to broaden the engine load range and optimize engine thermal efficiency. In the light of increasingly stringent diesel emission limits, changes in conventional combustion mode, addition of active elements to the fuel or after treatment of exhaust gases are needed. Concerning changes in combustion mode, even though considerable progress has been made in this research field over the last decade, the large scale adoption of HCCI (Homogeneous Charge Compression Ignition) diesel engines in commercial vehicles is currently not possible. The main challenges facing HCCI diesel engines are their limited operational range, lack of direct control of combustion phasing and increased HC and CO emissions, as reported in some applications [5,6]. Taking into account the addition of additives to the diesel fuel, cerium-based compounds have been used in order to lower PM emissions and to enhance oxidation rates but engine tests have shown an insensitive effect on gaseous pollutant emissions (HC, CO and NO_x).

Improvements in diesel engine designs will not be enough to meet the requirements of future legislation for both soot and NO_x. Therefore, after-treatment technologies are the object of intense research efforts. A variety of technologies have been proposed to abate NO_x and soot; the combination of filters and combustion catalysts for particle removal and catalytic traps for nitric oxide reduction are among the most extensively investigated alternatives.

The main concern for the application of filters to remove soot particles is the regeneration step. Both thermal and mechanical methods must be ruled out due to either their inefficiency or inapplicability. The most widely studied way to conduct the regeneration is using filters combined with a suitable catalyst, which enables simultaneous soot filtration and combustion [2]. With this system, the filter could be regenerated at a temperature comparable to that of diesel exhaust gases. Soot oxidation catalysts have been applied in three different ways: (i) as precursor compounds, soluble in the diesel fuel; (ii) as catalytic compounds that are injected upstream of the soot-loaded filter; and (iii) as a catalytic coating on the filter itself. Because of the logistic and possible environmental problems related to the first two options, the last option is the most desirable way to apply a catalyst for soot oxidation [7].

Another matter of concern is the presence of nitric oxide in diesel exhausts (typically around 1000-2000 ppm). This toxic compound should be eliminated in order to accomplish the increasingly severe limits imposed by law. NO_x adsorbers (traps), combined with a reduction catalyst, constitute a relatively new control technology, which has been developed for partial lean burn gasoline engines and for diesel engines [8]. The adsorbers, which are incorporated into the catalyst washcoat, chemically bind nitrogen oxides during lean engine operation. After the adsorber capacity is saturated, the system is regenerated and released NO_x is catalytically reduced during a period of rich engine operation [9]. In the case of diesel engines, since a rich operation is not feasible, periodic fuel injections are necessary.

On the other hand, nitric oxide favors the efficiency of the catalytic combustion of soot, because NO_2 is an oxidant stronger than O_2, enabling the regeneration of the filter at lower temperatures. This is the principle of CRT filter (Continuously Regenerating Technology). This system was developed and commercialized by Johnson Matthey, Catalysts division. It is claimed that the third generation CRT filter is able to reduce HC and PM by over 90% and CO by over 70%, but is not efficient to reduce NO_x emissions.

The CRT particulate filter utilizes Johnson Matthey's patented process which, as said above, oxidizes soot with NO_2 at a lower temperature than with oxygen [10]. This lower temperature is compatible with typical diesel exhaust temperatures, so no supplemental heat is required. The device is made up of two chambers. The first contains a substrate coated with a proprietary, highly-active platinum oxidation catalyst designed to oxidize a portion of the NO in the exhaust to NO_2. NO_2 generation is the key to the oxidation of soot collected by the wall flow filter and the heart of the Johnson Matthey's patent. The same catalyst also converts CO and HC into CO_2 and H_2O. In the second chamber, exhaust flows through a wall-flow filter, where gaseous components pass through, but the soot is trapped on the walls of the filter. The trapped soot is then removed by the NO_2 produced by the catalyst in the first chamber [11].

The complexity of the systems that are commercially available nowadays, motivate increasing efforts devoted to obtain more simple processes. In this vein, soot oxidation in catalytic traps has been recently proposed as an alternative to simultaneously eliminate both NO_x and particle contaminants [12]. In our group, it has been demonstrated that the reaction between soot particles and NO_x molecules originated by the decomposition of the trapped nitrate species is feasible when a catalytic trap composed of a mixed oxide of Ba, Co, K and Ce is used [13]. The reaction between nitrate species and soot is also the basis of the diesel particulate and NO_x reduction process (DPNR Toyota system). Nitrates are formed in a

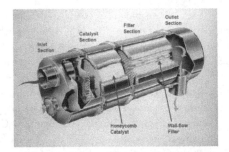

Figure 2. CRT particulate filter utilizes Johnson Matthey's.

catalytic trap composed of an alkali metal oxide after the NO to NO_x reaction takes place over the Pt atoms. It is believed that surface nitrates are decomposed producing NO and very reactive O atoms, which are responsible for soot oxidation [14,15]. The fate of NO molecules is to form nitrates again. Once it is saturated, the alkali-metal trap should be regenerated by running the engine in rich conditions or by adding fuel into the exhaust stream. However, in another work [16] it is reported that the presence of soot particles prevents NO_x traps from employing their full trapping capacity, which could result in a lack of efficiency of the DPNR Toyota technology.

3. Structured catalysts

A large number of catalyst formulations have been reported for soot combustion, and the soot-to-catalyst contact appears to be one of the most important problems to overcome. Some reviews have recently offered updates on new technologies for diesel emission control. Despite the abundant studies on both soot combustion and NO_x abatement carried out with powder formulations, only a handful of papers have been published using structured catalysts, which constitutes a more realistic approach. In particular, for soot particles abatement, an adequate morphology of the catalyst is required so as to improve the contact between catalyst and soot. For practical purposes, the catalytically active component should be supported as a film on a structured substrate, thus allowing simultaneous soot filtration and combustion.

Even though powder catalysts are unable to be used under real conditions in the exhaust pipes, the powder form is very useful to study the catalytic performance and the effect of different operating variables on their activities such as the soot-catalyst contact type [17-19], which is a very important factor in this type of reaction. Other factors of interest are the soot to catalyst ratio and the concentration of NO_x and O_2, because the concentration of these gases can change during the diesel engine operation [20].

A large number of powder catalysts have been investigated for controlling the diesel engines exhaust emission, the main objective of most of these studies being to improve the catalytic performance for soot combustion. There are numerous articles reporting the use of precious metals supported on Al_2O_3, Ce-Zr mixed oxide [21], CeO_2 and TiO_2 [18] for this application, although the cost of these metals is really high. For this reason, other bulk catalysts have been extensively studied in the last decade such as hydrotalcite-like compounds (CuMgAl [22] and $Mn_xMg_{3-x}AlO$ [23]), spinels ($ZnAl_2O_4$ [24] and $CoAl_2O_4$ [25]), perovskites ($LaMnO_3$ [26], $La_{1-x}K_xCo_{1-y}Ni_yO_{3-\delta}$ [27], $La_{0.9}K_{0.1}Co_{1-x}FexO_{3-\delta}$ [28] and $BaCoO_{3-y}$ [29]), mixed oxides (La-Ce oxide, Nd-Ce oxide, Fe-Ce oxide, Cu-Ce oxide [30]) and single oxides (SrO, CeO_2, Co_3O_4, La_2O_3 [31]). Besides, catalytic formulations containing oxide components supported on different oxides have resulted good catalysts for the control of diesel engines exhaust emissions. For example, Co_3O_4, KNO_3 and $BaCO_3$ supported on ZrO_2 (Co,Ba,K/ZrO_2) or CeO_2 (Co,Ba,K/CeO_2), where there occurs a combination of the moderated activities of ZrO_2 or CeO_2 with the high activity of dispersed Co_3O_4 for the soot combustion, along with the fact that KNO_3 favors the soot oxidation by improving the contact between the soot and the catalytic surface [13,32].

The main function of the catalysts mentioned above is that of being active for the catalytic combustion of soot, but other catalysts that have been developed for the abatement of NO_x emissions (catalytic traps) can also be employed for soot combustion. One of the most widely used compounds for the absorption of NO_x is barium, which traps this gaseous contaminant through the formation of the corresponding nitrate. At high temperatures, nitrate decomposes releasing NO_2, which in turn reacts with soot.

To reduce the NO_x concentration, the selective catalytic reduction (SCR) is the most commonly employed method, and NH_3 is the most common reducing agent. Commercially, NH_3 is generated in situ by urea decomposition and the BlueTec system developed by Mercedes Benz uses a 32.5% urea solution for NH_3 generation. The best SCR catalysts are zeolites exchanged with different metals as Fe or Cu [33,34], although oxides as CeO_2-TiO_2 or Ce-W-Ti mixed oxides have also been tested [35,36].

In the reaction of NH_3 with NO, O_2 is required to form H_2O so as to consume the excess hydrogen from NH_3. The reaction of NH_3 with NO_2 is a much slower reaction. However, equal amounts of NO and NO_2 undergo selective reduction with NH_3 in a reaction that is an order of magnitude faster than the reaction of NH_3 with either NO or NO_2 alone. This has been called "fast SCR" reaction, and O_2 does not take part in the reaction. As it can be noticed, NO_2 has the role of reacting with the hydrogen that O_2 consumes in the reaction of NO with NH_3. Since this mixed reaction is indeed "fast", it is important to adjust the NO/NO_2 ratio by the appropriate oxidation of NO to NO_2 before the SCR catalyst on a vehicle, and this is done by carefully selecting the catalyst [1].

The powder catalyst itself cannot be used for a practical application in the diesel engine exhaust pipe and as stated above, a catalytic coating on the filter appears to be the best option to use a catalyst for soot oxidation Therefore, a substrate containing the catalytic coating must be used as a system that can withstand the severe operating conditions of the exhaust pipe, high flows, high corrosive atmospheres, thermal shocks and vibrations. These substrates must have filtration properties, because one of the pollutants is solid (soot particles). The filter walls can be coated with a catalytic layer, in which the retained soot particles are burnt and the NO_x trapped at the temperature of the exhaust gases.

The most extensively used substrates for this purpose are ceramic monoliths [37] (Figure 3a), which consist of square parallel channels along the axial direction with thin porous walls. To act as a filter, the channels are open at one end, but plugged at the other one. Therefore, the exhaust gas soot particles are forced to flow through the filter walls, which are coated with a catalytic layer.

Other substrates used as filters are foams [38] (Figure 3b), which can be made of either ceramic or metallic materials. The rigid open-cell foam type allows exhaust gases to flow through and traps soot particles on its complex structure. The filter obstructions by soot particles is difficult since these soot particles are catalytically burnt through the catalytic layer coated onto the foam surface, i.e., foams provoke low pressure drops.

Although monoliths and foams are the most widely studied substrates for the preparation of structured catalysts, other merging substrates that can be considered for controlling the diesel engines exhaust pollutants are ceramic papers [39] (Figure 3c) and wire meshes [40] (Figure 3d). For the former, the catalytic components can be either coated onto the ceramic fiber or held as particles between the ceramic fibers [20] and for the latter, similar to foams, the filaments of the wire mesh can be coated with the catalyst and act as a catalytic filter [41].

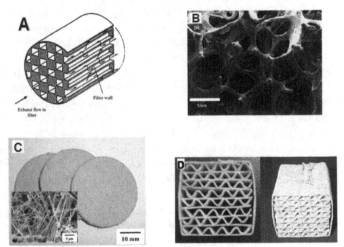

Figure 3. Four types of structures can be used to make a catalytic diesel particulate filter: A. Monolith, B. Ceramic Foam, C. Ceramic Paper and D. Wire Mesh.

Meille [39] described different coating methods: washcoating using a suspension (La_2O_3–Al_2O_3, Pd/ZnO, CuO/ZnO–Al_2O_3), sol–gel deposition (Pt/Al_2O_3, Rh/Al_2O_3, Pd/Al_2O_3, Pd/La_2O_3, Pd/SiO_2), hybrid method between suspension and sol–gel (CeO_2-ZrO_2-La_2O_3-Al_2O_3, CuO/ZnO-Al_2O_3), electrophoretic deposition (Al_2O_3), electrochemical deposition (Cu-Zn), electroless plating (Cu–Zn) and impregnation (Rh, Ni/La_2O_3, Fe_2O_3). Other methods described are chemical vapor deposition (Al_2O_3, Mo_2O_3), physical vapor deposition (Cu, Pt,

Mo, Zr, La$_2$O$_3$, Al$_2$O$_3$), flame assisted vapor deposition (NiO–Al$_2$O$_3$) and flame spray deposition (Au/TiO$_2$).

This chapter mainly focuses on different types of structured catalysts developed in our group for the control of diesel pollutants. Cobalt, barium and potassium deposited either on ZrO$_2$ or CeO$_2$ powders resulted efficient for the combustion of soot. These elements have been used to coat different foams and monolithic structures [42-43]. In order to evaluate the performance of these systems, soot particles were impregnated onto the structured catalysts using slurry of soot in n-hexane, thus giving place to loose contact. Despite the non-homogeneous nature of the foam, the application of the said method yielded satisfactory results regarding the reproducibility towards the catalytic behavior, comparable to that obtained with the powder catalyst.

3.1. Monoliths

The monolith catalysts were first developed in the 1970s for automotive exhaust gas treatment mainly to control hydrocarbon and CO emissions. After that, significant improvements were introduced until obtaining the three way gasoline catalyst converter (TWC). Operating under stoichiometric air to fuel ratio, this TWC simultaneously converts CO, HC and NO$_x$ into CO$_2$, H$_2$O and N$_2$. During the last three decades, monolith reactors have been widely used as air pollution control systems due to their relevant characteristics: low pressure drop, high specific external surface area and less catalyst attrition compared to those of the packed-bed reactors. Standing out among these applications, the one used by certain commercial airlines in order to ensure a clean and safe environment in the airplane cabin, namely the metal monolith system, can be mentioned. The light weight of a metal monolith coupled with a low pressure drop makes this a cost-effective technology. In some industrial processes, nitrogen oxides (NO$_x$) are produced and the structured catalysts made with monoliths are used for the selective reduction of NO$_x$, with NH$_3$. Besides, a large amount of volatile organic compounds (VOCs) is produced in several industries. In these cases, the monolith catalyst is used to remove the VOCs. After many years of research and development, the catalytic combustor has now been commercialized for gas turbines replacing traditional burners. It uses large air excess, generates sufficient temperatures to operate the turbine, but with virtually no emission of CO, HC or NO$_x$. Other applications are the hydrogen generation for the fuel cell by steam reforming of hydrocarbons, water gas shift and preferential oxidation of CO [44].

Concerning diesel exhaust particulate removal, wall-flow diesel particulate filters (DPFs) are considered the most effective devices for the control of diesel particulate emissions. A requirement for the reliable operation of the DPFs, however, is the periodic and/or continuous regeneration of the filters.

This type of substrate was developed for the catalytic after-treatment of gasoline engine exhausts. Some modifications were made to the traditional monolith to be used as diesel particulate filter. The monoliths most widely used are made of cordierite (ceramic material consisting of magnesia, silica, and alumina in the ratio of 2:5:2) or silicon carbide, materials that can withstand high temperatures, as those achieved during the regeneration of the diesel particulate filter. Besides, these substrates are also made of aluminum titanate, alumina or mullite. Metallic monoliths, extensively used as structured supports of catalysts in other reactions, are not porous and therefore not used as diesel particulate filters.

Figure 4. Fabrication of a ceramic monolith by extrusion.

The ceramic monolith structures are produced via extrusion [45], as shown in Figure 4. The monolith cells can have different geometries; the most popular is the square one, although there are monoliths with hexagonal and triangular cells (Figure 5). The characteristic parameters that describe the monolith structure are presented in Table 1 [46]. The cell density is defined as the number of cells per unit area, where the area is generally described in square inches and the most common monoliths have 100, 200 or 400 cpsi (cells per square inch). Since the catalytic filters have the double function of filtering particles and burning them, the structural properties (porosity, surface area, wall thickness) have a central role in their efficiency. The wall thickness is an important parameter since it is related to the filter capacity and pressure drop caused by the monolith. Although high porosities are desirable for wall flow-through type filters, this make structures more fragile and implies low rupture modulus values.

As said above, there are different methods to generate a catalytic layer onto the substrate walls. In general, the combination of the active components for the layer is chosen from those active catalysts previously evaluated as powders. In what follows, a summary is included of some methods employed for the synthesis of structured catalysts using a monolith as structured substrate:

Figure 5. Different forms of cells in a monolith.

Characteristics	Commercial denomination		
	100/17	200/12	400/4.5
Cell Density			
(cpsi)	100	200	400
(cpscm)	15.5	31.0	62
Wall Thickness			
(in)	0.017	0.012	0.0045
(mm)	0.432	0.302	0.102
Open Frontal Area (OFA)			
(%)	34.5	34.5	84
Geometric Surface Area (GSA)			
(in^2/in^3)	33.3	47.0	69
(m^2/l)	1.31	1.85	2.72
Hydraulic Diameter			
(in)	0.0813	0.059	0.0455
(mm)	2.11	1.49	1.09
Modulus of Rupture			
(psi/cpscm)	350	300	
Porosity			
(%)	48	48	48
Coefficient of Thermal Expansion			
($\times 10^{-7}$ cm/cm/°C(25-800°C))	5	5	5
Maximun average			
Mean Pore Size			
(micron)	13	13	13

Table 1. Characteristics of three commercialized ceramic monoliths

Solution combustion synthesis: This method consists in the dipping of the ceramic support in the aqueous solution of its precursors, which is then placed into an oven at a high temperature. The aqueous phase is rapidly brought to boil, the precursors mixture ignite and the synthesis take place in situ. This method has been applied to deposit catalysts as perovskites or spinels, as $La_{0.9}K_{0.1}Cr_{0.9}O_3$ [47,48], $LiCoO_2$ [49,50] and $LaCr_{0.9}O_3$ [51] over ceramic monoliths.

Impregnation: The catalysts are directly deposited on the monolith by immersing the structured substrate into a solution of the precursors of the catalyst; then the monolith is blown to eliminate the excess of solution, and finally it is dried and calcined. This method was used to deposit Cu/K/Mo over the monolith [7,52]. Also, coatings of KNO_3/CeO_2, K_2CO_3/CeO_2 or KOH/CeO_2 were performed over monoliths. In a first step, Ce was deposited using a solution of $Ce(NO_3)_3$ and once CeO_2 was formed after calcination, potassium precursors were deposited [53].

Washcoating: This method consists in immersing the monolith into a suspension of either catalyst particles or catalytic precursors; in the former case, the catalyst is synthesized as powder and then, the suspension is prepared. With this method Co,K/Al$_2$O$_3$ or Cu,K/Al$_2$O$_3$ [54] and Cu–K/Al$_2$O$_3$, Co–K/Al$_2$O$_3$ or V–K/Al$_2$O$_3$ [54] were deposited over monolithic structures.

Combination methods: In some cases, the combination of more than one technique is used. For example, Banús et al. deposited the catalyst Co,Ba,K/ZrO$_2$ over monoliths in two steps. First, a layer of ZrO$_2$ was deposited on the monolith walls by the washcoating method, using a ZrO$_2$ colloidal suspension and then, after calcination, the active metals (Co, Ba and K) were impregnated [42].

3.2. Foams

Among the various diesel particulate filters (DPFs), ceramic foams are attractive structured systems that can be prepared from a range of materials and have characteristics that make them desirable as substrates for structured heterogeneous catalysts. They exhibit high porosities with a significant degree of interconnectivity among spherical-like cells through openings or windows, which results in low pressure drop. While the more conventional wall-flow type monoliths act as "cake filters", foams act as deep bed filters. In monoliths, the surface filtration ("cake formation") gives good particulate collection efficiency, but a significant pressure drop occurs as the layer of soot particulates deposited on the filter walls grows. In the case of the deep filtration type filters (foams), a good penetration/dispersion of particulate inside the trap matrix is obtained, thus allowing constant collection efficiency. Although the retention efficiency is often low, this can be overcome with an appropriate trap design. The use of different types of foams has been the object of several interesting research studies aimed at eliminating diesel contaminants produced by diesel vehicles. Other catalytic applications, such as CO$_2$ reforming, have been explored. A low surface area ceramic substrate gives desirable physical and mechanical properties but the choice of an adequate washcoat provides a high surface area so that catalytic films could enhance their activities when depositing them on the washcoat-structured substrate.

There are two types of foams, those with open pores [56] (Figure 6a) and those with closed pores [57] (Figure 5b). Nevertheless, only the former type can be used for making structured catalysts since the latter, containing closed pores, make it impossible for any gas to flow through and is only used as isolating material. Foams are mainly defined by the pore number per linear inch (pores per inch, ppi). In Figure 7, open pore metallic foams with different ppi are shown [58]; the higher the ppi number, the smaller the pore dimensions. Smaller pore size increases the pressure drop, as observed in Figure 8 [59]. The filter efficiency of foams is related to a high ppi number and to the filter length. On the other hand, a diesel particulate filter with a high ppi number provokes a high pressure drop, which could deteriorate the performance of diesel engines.

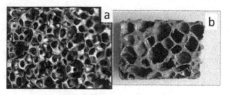

Figure 6. Type of foam. Open pores (a) and closed pores (b).

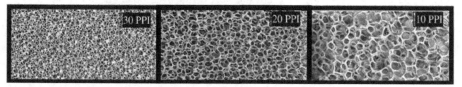

Figure 7. Metallic foam of different pores per inch.

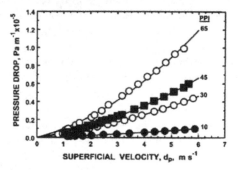

Figure 8. Pressure drop as a function of pores per inch (ppi).

Ceramic or metallic foams of open pores can also be used as substrate for a catalytic coating in numerous reactions or catalytic processes such as methane catalytic oxidation, hydrogen production, engine gas exhaust purification, ammonia oxidation to obtain nitric acid, methanol partial oxidation, hydrocarbon total oxidation, CO_2 reforming, and Fischer–Tropsch synthesis, among others.

Metallic foams are made of either a single metal such as aluminum, nickel, or copper, or of alloys such as brass, Ni–Fe, Ni–Cr or Ni–Cr–Al, Fe–Cr–Al or stainless steels, whereas ceramic foams are made of Al_2O_3, $Mg_2Al_4Si_5O_{18}$, SiC, Al_2O_3-ZrO_2, Y_2O_3-Al_2O_3, Al_2O_3-SiO_2 and TiO_2. By far, the most studied ceramic foams are the Al_2O_3 ones [60].

Different techniques are used for the manufacturing of ceramic foams:

- Replication of artificial or natural foams [61,62],
- Employing generator agents of pores, as vegetal flours or polymers [63,64] and
- Foaming methods, where either by agitation [65] or by the fast evaporation of the solvent, pores are formed [66].

On the other hand, metallic foams are produced in different ways such as [67]:

- Bubbling gas through a molten metal.
- Stirring a foaming agent (typically TiH_2) into a molten alloy (typically an aluminum alloy) and controlling the pressure while cooling.
- Consolidating a metallic powder (the most commonly used ones are aluminum alloys) with a particulate foaming agent (TiH_2) followed by heating into the mushy state. When the foaming agent generates hydrogen, the material is expanded and the foam is formed.
- Manufacturing ceramic mold from a wax or polymer-foam precursor, followed by burning-out of the precursor and pressure infiltration with molten metal or metal powder slurry which is then sintered.
- Vapor phase deposition or electrodeposition of a metal onto a polymeric foam precursor, which is burned out, leaving cell edges with hollow cores.
- Dissolution of a gas (typically hydrogen) in a liquid metal under pressure, allowing it to be released in a controlled way during subsequent solidification.

Concerning particulate filters, as stated, the most commonly used foams are ceramic ones, made of α-Al_2O_3 and 65 or 50 ppi. According to the process used to produce the foam, the final product is between 1 or 2 inch thickness, so that for the fabrication of the diesel particulate filter, several foam discs are placed in a container [68], as shown in Figure 9. Besides the typical material used for diesel particulate foam filters (α-Al_2O_3), other materials are SiC, zirconia toughened alumina (ZTA) and zirconia toughened mullite (ZTM). These materials are selected due to their resistance to corrosion in atmospheres and high temperatures, which are operating conditions of diesel exhaust filters. Metallic foams without any pre-treatment are not suitable to be used as filters because of the low resistance under severe operating conditions. However, after either passivation or coating with the appropriate compounds, these metallic foams can be used as diesel particulate filter. Only a few articles concerning metallic foams as filters are available in the literature and they mainly report the use of foams made of stainless steels, FeCrAl alloys and nickel.

Ciambelli and coworkers have coated the ceramic foam walls with Cu/V/K/Cl, through several cycles of immersion into an aqueous solution of the catalyst precursor salts, drying at 120°C and calcining at 700°C overnight [68-72]. Through this process, different ceramic foams have been coated with vanadates [73-77] or with Pb-Co [78]. Another method for obtaining the catalytic coating is by dipping the ceramic foam into a molten solution containing precursor salts and then, eliminating the solution excess by blowing [79-81] and finally, calcining the coated monoliths.

Banús et al. reported that $Co,Ba,K/ZrO_2$ is a good powder catalyst for both soot combustion and NO_x trapping [32] and they produced $Co,Ba,K/ZrO_2$ coatings onto calcined metallic foams [43] and ceramic foams [61]. The first step for obtaining this catalytic coating was to deposit a ZrO_2 layer by immersing the foam into a commercial suspension of ZrO_2 nanoparticles (ZrO_2/foam). Then, the impregnation of the active metals (Co,Ba,K) to the zirconia layer was done by dipping the ZrO_2/foam into a solution of the metal salts. The

Figure 9. Image of radial flow foam trap.

ZrO₂ layer increases the surface area and in the case of the metallic foam, it also protects its surface walls from the corrosion and the high temperature. It has been reported that CeO₂ also has the same effect over FeCrAl alloy fibers [82]. In the other case, pure nickel foam is coated with a powder FeNiCrAl alloy, and after the rapid transient liquid phase sintering, a homogeneous alloy foam is formed [83].

3.3. Wire-meshes

It has been reported that by covering a wire mesh with a porous ceramic layer through a modified thermal spray technique an active and durable catalyst can be produced [37]. The surface area of the porous ceramic layer can be increased by an in-situ precipitation technique, through washcoating or by means of sol-treatments. The ceramic layer is finally covered with a catalytically active material, e.g., Pd, Pt or metal oxides. Wire-mesh catalysts combine the excellent mass and heat transfer performance of a pellet-type catalyst with the low pressure drop of a monolith. The effects of pore diffusion during combustion are relatively small due to the shell like design of the catalyst layer. Thus, the catalyst efficiency is considerably improved compared with pellet or monolith catalysts. The effects of catalyst clogging and fouling can be handled as these structured catalysts are easy to disassemble and to clean. Wire-mesh catalysts also offer a great flexibility with respect to dimensions that make retrofit installations easier.

A study of soot combustion on wire meshes formed inside a cylindrical filter is being carried out in a joint collaborative work between our group and a team from the University of the Bask Country. Preliminary results using an active phase composed by Co,Ba,K/CeO₂ coated on stainless steel wire meshes are promising.

3.4. Ceramic papers

A material that has been extensively used as particle filter is cellulose paper, which can be flexibly adapted to different conformations. Besides, the paper-like structure, with interconnected pore spaces formed by the fiber network, may provide an adequate matrix that promotes desirable exhaust gas diffusion within the catalyst layer where filtration of particulate matter along with the combustion of trapped soot would take place. However, in the case of diesel soot oxidation, the filter should resist the high temperatures of the exhaust, a requirement that cannot be met by cellulose fibers. It has been recently shown that the use of ceramic fibers along with cellulose ones and the application of papermaking techniques result in high-temperature resistant ceramic papers. The addition of cellulosic fibers in low proportion favors the fiber mat consolidation, not only when the mat is wet but also when the mat is dried. And finally, a calcination step is needed, by which cellulose fibers are burnt and a matrix composed of ceramic fibers is obtained. In this way, a stable porous structure suitable for gas flow-through applications can be obtained.

Different types of ceramic fibers can be used thus allowing the preparation of a variety of structured matrices. However, fiber length constitutes a limitation and ceramic fibers of at least 600 μm of length are needed. Most generally used ceramic fibers are a mixture of SiO_2 (50%) and Al_2O_3 (48%) and they are frequently denoted only as ceramic fibers, although some papers report the use of fiber rods of mullite ($3Al_2O_3.2SiO_2$) [84] or glass fiber [85].

Likewise, the addition of a binder is necessary to obtain optimum mechanical properties that allow an easy manipulation of ceramic papers after calcination, i.e., after cellulose fibers are burnt. The kind and amount of ceramic binder should be selected in order to obtain a strong but flexible structure. In general, most published papers employ alumina sol as a binder for enhancing the physical strength of the catalyst paper [38, 84, 86-89], although some papers use TiO_2 sol [90,91] or SiO_2 sol [92]. Our group is currently studying the use of borate compounds as binder to enhance the mechanical properties of ceramic papers. It has been found that ceramic papers with different properties can be obtained by modifying the final calcination temperature. Although none of the borate tested melt after calcination up to 750°C, individual particles began to sinter as the calcination temperature increased thus joining ceramic fibers and giving ceramic papers better mechanical properties (unpublished results).

Either during or after the synthesis of the ceramic paper, a catalyst can be added; thus, a ceramic paper with catalytic properties is obtained. In the last few years, the idea behind the so called "catalytic paper" has been to apply this paper for making structured catalysts for different reactions, for example methanol steam reforming [90-91,93-94] and the reduction of NO_x in exhaust gases [85,92]. As said before, diesel soot oxidation in NO_x catalytic traps has been recently proposed as an alternative to simultaneously eliminate NO_x and particles [95]. Thus, the ceramic catalytic papers could also be applied as structured systems, whose flexibility make them adaptable to different conformations and geometries.

The incorporation of catalysts (single and mixed oxides and zeolites) either during paper manufacturing or after the pressed mat was obtained was studied. The ceramic paper was prepared by a papermaking technique with a dual polyelectrolyte retention system. Considering that wood fibers and fines in an aqueous system are negatively charged due to the ionization of functional groups associated with hemicelluloses, oxidized cellulose and oxidized lignin [85], this method enhances fiber retention. But most important is also the effect of polyelectrolytes in retaining ceramic fibers, catalytic particles and the binder. Different cationic and anionic polymers can be used, the most common ones being PolyDADMAC (cationic polydiallyldimethylammonium chloride), PVAm (cationic polyvinyl amine), as cationic polymers and PES-Na (sodium-polyethylenesulphate), C-PAM (cationic polyacrylamide) and A-PAM (anionic polyacrylamide) as anionic polymers.

A ceramic paper to the diesel soot combustion reaction in air, made of ceramic fibers, cerium oxide and potassium nitrate was prepared. The synthesis of this material was easy and fast, and provided a high-temperature resistant paper. The presence of potassium and cerium as active ingredients conferred high catalytic activity for the reaction studied under loose contact, showing a maximum combustion rate at 395°C [20].

The results show that the application of the catalytic paper concept is a promising alternative for making catalytic filters to be used in the abatement of diesel contaminants. The fibrous nature of the paper would allow modulating the porosity in order to improve the filtering capacity. The synthesis of the catalytic ceramic paper is easy, fast, and provides a material that can be simply adapted to different conformations and geometries. The presence of potassium and cerium as active ingredients confers a high catalytic activity for the diesel soot combustion reaction under loose contact.

The papermaking technique has also been applied for the preparation of both cellulosic and ceramic papers containing NaY zeolite, for different uses: those that only contained cellulosic fibers for low temperature applications (cellulosic papers) and those prepared using both cellulosic and ceramic fibers (ceramic papers) for high temperature applications. The first preparations of these zeolitic papers were performed without the addition of binder, and although a decrease in the mechanical properties after calcination was observed, zeolitic ceramic papers resulted easy to handle for practical applications [96]. SEM (Scanning Electron Microscopy) images indicated a good dispersion of zeolite particles within the cellulosic paper whereas in the ceramic paper they appeared anchored on ceramic fibers.

The NaY zeolite resulted homogeneously distributed and the zeolitic cellulosic papers were mechanically acceptable, which makes them promising for applications at low temperatures. In this line, the zeolitic papers prepared resulted efficient as toluene sorbents, the adsorption capacities of zeolitic ceramic papers being higher than those of zeolitic cellulosic papers. Toluene adsorption values demonstrate that zeolite dispersed into the cellulose/ceramic matrix is as effective as powder massive zeolite in retaining the hydrocarbon, which highlights the potential application of these zeolitic structures as sorbent materials both for low and high temperatures. Although a decrease in their mechanical properties was observed due to calcination, zeolitic ceramic papers resulted easy to handle for practical applications.

The purpose of this work is to explore the possibility of using catalytic papers to make efficient self-regenerating diesel soot filters. To this end, different binder components are being employed to enhance the flexibility of ceramic papers; and the amount of ingredients used during the papermaking process is being modified. In this way, paper thickness, catalyst amount, etc, can be varied. Besides, the use of zeolitic ingredients added to the paper constitutes a promising way to generate new and effective SCR structured catalysts. Moreover, the intercalation of paper beds containing different catalysts (sandwich catalysts) could be used for the one-step process to eliminate different contaminants.

4. Catalytic tests of powder and structured catalysts

Peralta et al. studied the optimal parameters for evaluating powder catalysts for the catalytic combustion of soot and they concluded that the best conditions were: a catalyst:soot mass ratio of 20:1 (a mixing time of 6 min), a flow rate of 40 ml/min, when loading 10 mg of sample and the oxygen partial pressure was 0.05 bar [97]. The catalysts were evaluated with real soot obtained by diesel combustion in a glass vessel [20] or different materials substitutive of the soot, as carbon black [98] or commercial soot (i.e.: 390127-25G of Aldrich) [99].

To evaluate the performance of structured catalysts, more than one method can be applied. One option is the incorporation of the soot using a suspension of soot in a solvent, for example soot in n-hexane [20,35,52] or n-heptane [71,76]. In these studies, the soot concentration to incorporate different amounts of soot in the structured catalysts is varied and thus the performance of these structures with different amounts of soot is evaluated. Another option consists in generating a gaseous stream containing soot particles, which can be generated from the burning of diesel fuel [60,70] or by generating a flame from the burning of ethane [100] or acetylene [73] as fuels. On the other hand, previously generated soot particles can be directly dispersed in a gaseous stream, thus generating an aerosol [26,75,79,101].

The filter should be able to trap and burn soot particles as it collects them, so that the filter does not saturate. This is another way of evaluating catalytic activity: to determine the autoregeneration performance of the structured catalysts. This consists in the evaluation of the capacity of burning soot particles from a diesel exhaust as they are trapped in the filter. There are several ways of simulating the diesel exhaust: one consists in collecting one portion of the diesel engine exhausts and to use it to evaluate the performance of the structured catalyst [72,102,103] or simply placing the filter containing the structured catalyst in the diesel engine exhaust [104].

Other variables that can be studied are O_2 and NO concentrations, because these concentrations may change during the operation of the diesel engines and the concentration of these gases is easily varied in the laboratory in an experiment of temperature programmed oxidation.

Another variable that is being studied is the contact between the soot particle and the catalyst. For powder catalysts, mixtures of soot and the catalyst can be prepared by producing either tight or loose contact. In order to check the intrinsic activity of a catalyst so as to determine the best performance of the formulations, tight conditions are preferred [19]. Nevertheless, loose contact represents real conditions so that for structured catalysts, which possess technological applicability, loose contact is preferred to check the performance of the system [52,73].

As structured catalysts can be applied in devices for their use in cars, the adherence of the catalytic coating is fundamental [43]. This stability is studied subjecting the structured catalyst to ultrasonic treatments in different solvents, as acetone [40,52].

The measurement of the pressure drop during the operation of the diesel particulate filter constitutes a matter of concern, since the catalytic filter should be able to burn soot as it is trapped, so as not to plug the filter. For this purpose, pressure drop versus time is measured and, if the pressure drop increases constantly, the filter cannot burn the trapped soot and finally it becomes obstructed. These measurements depend on the temperature of operation of the filter so that the catalyst should exhibit its best performance in the temperature range of the diesel engine exhausts.

Fino et al. evaluated two catalysts: $LaCr_{0.9}O_3$ and $Cs_2O.V_2O_5$. When these catalysts were tested as powders, they showed a good performance for the catalytic combustion of soot. However, when these catalysts were coated over foam, only the $Cs_2O.V_2O_5$ catalyst had an acceptable performance, while the $LaCr_{0.9}O_3$ catalyst showed a worse performance than that of the uncoated foam. These behaviors were manifested when the structured catalysts were studied under real conditions. In this study, the pressure drop of the foam coated with $LaCr_{0.9}O_3$ was higher than that observed for the uncoated foam. This indicates that although a system could present good activity as a powder, tests using the structured systems are necessary to discriminate between different formulations [47].

5. Conclusions

It can be seen that nowadays the most popular systems for soot elimination in diesel exhausts are a complex combination of different types of physical and chemical processes. More compact processes are highly desirable to reduce costs and improve efficiency. To this end, the synergism between better understandings of the fundamentals of the catalytic reactions together with the development of new structured systems able to efficiently retain soot particles is highly desirable. It has been shown that an adequate morphology of the catalyst is required so as to improve the contact between the catalyst and soot and, simultaneously, the catalyst should be supported as a film on a porous structured substrate, thus allowing simultaneous soot filtration and combustion.

Although monoliths and foams are the most widely studied substrates for the preparation of structured catalysts, other merging substrates that can be considered for controlling the diesel engines exhaust pollutants are ceramic papers and wire meshes. In our laboratory, research works are ongoing, with the aim of developing these types of novel structures in order to make them efficient and economic diesel particulate catalytic filters. In this vein, papers made of ceramic fibers, cerium oxide and potassium nitrate were prepared which provided a high-temperature resistant paper with good activity for soot combustion at temperatures as low as 390°C. Moreover, the use of zeolitic ingredients added to the paper constituted a new way to generate new and effective SCR structured catalysts. Thus, the intercalation of paper beds containing different catalysts (sandwich catalysts) could be used for the one-step process to simultaneously eliminate soot particles and NO_x.

On the other hand, we have developed filters made of AISI-316 stacked wire meshes coated with the $Co,Ba,K/CeO_2$ catalyst, that showed good activity for the combustion of diesel soot. Wire-mesh catalysts combine excellent mass and heat transfer performance with low pressure drop, the catalyst efficiency being considerably improved compared with pellet or monolith catalysts. The effects of catalyst clogging and fouling can be handled as these structured catalysts are easy to disassemble and to clean. Wire-mesh catalysts also offer a great flexibility with respect to dimensions that make retrofit installations easier.

Author details

E.D. Banús, M.A. Ulla, E.E. Miró and V.G. Milt[*]
Instituto de Investigaciones en Catálisis y Petroquímica, INCAPE (FIQ, UNL – CONICET), Santiago del Estero, Santa Fe, Argentina

Acknowledgement

The authors wish to acknowledge the financial support received from ANPCyT, CONICET and UNL. Thanks are also given to Elsa Grimaldi for the English language editing.

6. References

[1] Twigg M.V. (2011) Catalytic control of emissions from cars. Catal. Today 163: 33–41.

[2] Fino D. (2007) Diesel emission control: Catalytic filters for particulate removal. Sci. Technol. Adv. Mater. 8: 93–100.

[3] Tree D.R., Svensson K.I., (2007) Soot processes in compression ignition engines. Prog. Energy Combust. Sci 33: 272-309.

[4] Gómez-García M. A., Pitchon V., Kiennemann A. (2005) Pollution by nitrogen oxides: an approach to NOx abatement by using sorbing catalytic materials. Environ. Int. 31: 445– 467.

[5] Lu X., Han D., Huang Z. (2011) Fuel design and management for the control of advanced compression-ignition combustion modes. Prog. Energy. Combust. Sci. 37 (2011) 741-783.

[6] Gan S., Ng H.K., Pang K.M. (2011) Homogeneous Charge Compression Ignition (HCCI) combustion: Implementation and effects on pollutants in direct injection diesel engines. Appl. Energ. 88 (2011) 559–567.

[7] Neeft J.P.A., Schipper, W., Mul, G., Makkee M. Moulijn, J.A. (1997) Feasibility study towards a Cu/K/Mo/(Cl) soot oxidation catalyst for application in diesel exhaust gases. Appl. Catal. B: Environmental 11, 365.

[8] Twigg M.V. (2007) Progress and future challenges in controlling automotive exhaust gas emissions. Appl. Catal. B: Environmental 70, 2-15.

[9] Heck R.M., Farrauto R.J. (2001) Automobile exhaust catalysts. Appl. Catal. A: General, 221, 443-457.

[10] http://www1.eere.energy.gov/vehiclesandfuels/pdfs/deer_2002/session11/2002_deer_ch atterjee1.pdf

---

* Corresponding Author

[11] Stanmore B.R., Tschamber V., Brilhac, J.F. (2007) Oxidation of carbon by NO_x, with particular reference to NO_2 and N_2O. Fuel, 87: 131-146.

[12] Krishna K., Makkee, M. (2006) Soot oxidation over NO_x storage catalysts: Activity and deactivation. Catal. Today 114: 48-56.

[13] Milt V.G., Querini, C.A., Miró, E.E., Ulla, M.A. (2003) Abatement of diesel exhaust pollutants: NOx adsorption on $Co,Ba,K/CeO_2$ catalysts. J. Catal. 220: 424-432.

[14] Nakatani, K., Hirota, S., Takeshima, S., Itoh, K., Tanaka, T. (2002) Simultaneous PM and NO_x Reduction System for Diesel Engines. SAE paper 2002-01-0957.

[15] Itoh, K., Tanaka, T., Hirota, S., Asanuma, T., Kimura, K., Nakatani, K. (2003) Exhaust purifying method and apparatus of an internal combustion engines. US patent, US 6,594,991.

[16] Sullivan J.A., Keane O., Cassidy A. (2007) Beneficial and problematic interactions between NO_x trapping materials and carbonaceous particulate matter. Appl. Catal. B: Environmental 75: 102-106

[17] Peralta M.A., Zanuttini M.S., Querini C.A. (2011) Activity and stability of $BaKCo/CeO_2$ catalysts for diesel soot oxidation. Appl. Catal. B: Environmental 110: 90– 98.

[18] Lim C.B., Kusaba H., Einaga H., Teraoka Y. (2011) Catalytic performance of supported precious metal catalysts for the combustion of diesel particulate matter. Catal. Today 175: 106–111.

[19] Sun M., Wang L., Feng B., Zhang Z., Lu G., Guo Y. (2011) The role of potassium in K/Co_3O_4 for soot combustion under loose contact. Catal. Today 175: 100-105.

[20] Banús E.D., Ulla M.A., Galván M.V., Zanuttini M.A., Milt V.G., Miró E.E. (2010) Catalytic ceramic paper for the combustion of diesel soot. Catal. Comm. 12: 46–49.

[21] Azambre B., Collura S., Darcy P., Trichard J.M., Da Costa P., García-García A., Bueno-López A. (2011) Effects of a $Pt/Ce_{0.68}Zr_{0.32}O_2$ catalyst and NO_2 on the kinetics of diesel soot oxidation from thermogravimetric analyses. Fuel Process Technol 92: 363–371.

[22] Wang Z., Li Q., Wang L., Shangguan W. (2012) Simultaneous catalytic removal of NO_x and soot particulates over CuMgAl hydrotalcites derived mixed metal oxides. Appl. Clay Sci. 55: 125-130.

[23] Li Q. ,Meng M., Xian H., Tsubaki N., Li X., Xie Y., Hu T., Zhang J. (2010) Hydrotalcite-Derived $Mn_xMg_{3-x}AlO$ Catalysts Used for Soot Combustion, NO_x Storage and Simultaneous Soot-NO_x Removal. Environ. Sci. Technol. 44: 4747–4752.

[24] Zawadzki M., Staszak W., López-Suárez F.E., Illán-Gómez M.J., Bueno-López, A. (2009) Preparation, characterisation and catalytic performance for soot oxidation of copper-containing $ZnAl_2O_4$ spinels. Appl. Catal. A: General 371: 92–98.

[25] Zawadzki M., Walerczyk W., López-Suárez F.E., Illán-Gómez M.J., Bueno-López A. (2011) $CoAl_2O_4$ spinel catalyst for soot combustion with NO_x/O_2. Catal. Commun. 12: 1238–1241.

[26] Li S., Kato R., Wang Q., Yamanaka T., Takeguchi T., Ueda W. (2010) Soot trapping and combustion on nanofibrous perovskite $LaMnO_3$ catalysts under a continuous flow of soot. Appl. Catal. B: Environmental 93: 383–386.

[27] Li Z., Meng M., Dai F., Hu T., Xie Y., Zhang J. (2012) Performance of K and Ni substituted $La_{1-x}K_xCo_{1-y}Ni_yO_{3-\delta}$ perovskite catalysts used for soot combustion, NO_x storage and simultaneous NO_x-soot removal. Fuel 93: 606-610.

[28] Li Z., Meng M., Li Q., Xie Y., Hu T., Zhang J. (2010) Fe-substituted nanometric $La_{0.9}K_{0.1}Co_{1-x}Fe_xO_{3-\delta}$ perovskite catalysts used for soot combustion, NO_x storage and simultaneous catalytic removal of soot and NO_x. Chem. Eng. J. 164: 98–105.

[29] Milt V.G., Ulla M.A., Miró E.E. (2005) NO_x trapping and soot combustion on $BaCoO_{3-y}$ perovskite: LRS and FTIR characterization. Appl. Catal. B: Environmental 57: 13-21.

[30] Muroyama H., Hano S., Matsui T., Eguchi K. (2010) Catalytic soot combustion over CeO_2-based oxides. Catal. Today 153: 133–135.

[31] Zhang R., Luo N., Chen B., Kaliaguine S. (2010) Soot Combustion over Lanthanum Cobaltites and Related Oxides for Diesel Exhaust Treatment. Energy Fuels 24: 3719–3726.

[32] Milt V.G., Banús E.D., Ulla M.A., Miró E.E. (2008) Soot combustion and NO_x adsorption on $Co,Ba,K/ZrO_2$. Catal Today 133–135: 435-440.

[33] Colombo M., Nova I., Tronconi E., Schmeißer V., Bandl-Konradb B., Zimmermann L. (2012) $NO/NO_2/N_2O–NH_3$ SCR reactions over a commercial Fe-zeolite catalyst for diesel exhaust aftertreatment: Intrinsic kinetics and monolith converter modeling. Appl. Catal. B: Environmental 111–112: 106–118.

[34] Colombo M., Nova I., Tronconi E. (2012) NO_2 adsorption on Fe- and Cu-zeolite catalysts: The effect of the catalyst red–ox state. Appl. Catal. B: Environmental 111–112: 433– 444.

[35] Shan W., Liu F., He H., Shi X., Zhang C. (2012) An environmentally-benign CeO_2-TiO_2 catalyst for the selective catalytic reduction of NO_x with NH_3 in simulated diesel exhaust. Catal. Today, in press.

[36] Shan W., Liu F., He H., Shi X., Zhang C. (2012) A superior Ce-W-Ti mixed oxide catalyst for the selective catalytic reduction of NO_x with NH_3. Appl. Catal. B: Environmental 115–116: 100–106.

[37] Koga H., Kitaoka T. (2011) One-step synthesis of gold nanocatalysts on a microstructured paper matrix for the reduction of 4-nitrophenol. Chem. Eng. J. 168: 420–425.

[38] Meille V. (2006) Review on methods to deposit catalysts on structured surfaces. Appl. Catal. A: General 315: 1–17.

[39] Konstandopoulosa A.G., Kostogloua M. (2000) Reciprocating flow regeneration of soot filters. Comb. Flame, 121: 488-500.

[40] Yang K.S., Mul G., Choi J.S., Moulijn J.A., Chung J.S. (2006) Development of TiO_2/Ti wire-mesh honeycomb for catalytic combustion of ethyl acetate in air. Appl. Catal. A: General 313: 86–93.

[41] Ahlström-Silversand A.F., Ingemar Odenbrand C.U., (1997) Thermally sprayed wire-mesh catalysts for the purification of flue gases from small-scale combustion of bio-fuel Catalyst preparation and activity studies. Appl. Catal. A: General 153: 177-201.

[42] Banús E.D., Milt V.G., Miró E.E., Ulla M. A. (2012) Catalytic coating synthesized onto cordierite monolith walls. Its application to Diesel soot combustion. Appl. Catalysis B: Environmental, submitted.

[43] Banús E.D., Milt V.G., Miró E.E., Ulla M.A. (2010) Co,Ba,K/ZrO2 coated onto metallic foam (AISI 314) as a structured catalyst for soot combustion: Coating preparation and characterization. Appl. Catal. A: General 379: 95–104.

[44] Heck R.M., Gulati S., Farrauto R.J. (2001) The application of monoliths for gas phase catalytic reactions. Chem. Eng. J. 82: 149–156.

[45] http://www.ikts.fraunhofer.de/en/Images/WabeExtruder_2_tcm244-32344.JPG.

[46] http://www.corning.com/WorkArea/showcontent.aspx?id=6465.

[47] Biamino S., Fino P., Fino D., Russo N., Badini C. (2005) Catalyzed traps for diesel soot abatement: In situ processing and deposition of perovskite catalyst. Appl. Catal. B: Environmental 61: 297–305.

[48] Cauda E., Fino D., Saracco G., Specchia V. (2004) Nanosized Pt-perovskite catalyst for the regeneration of a wall-flow filter for soot removal from diesel exhaust gases. Top. Catal. 30/31: 299-303.

[49] Fino D., Cauda E., Mescia D., Russo, Saracco G., Specchia V. (2007) LiCoO2 catalyst for diesel particulate abatement. Catal. Today 119: 257–261.

[50] Cauda E., Mescia D., Fino D., Saracco G., Specchia V. (2005) Diesel Particulate Filtration and Combustion in a Wall-Flow Trap Hosting a LiCrO2 Catalyst. Ind. Eng. Chem. Res. 44: 9549-9555.

[51] Fino D., Fino P., Saracco G., Specchia V. (2003) Innovative means for the catalytic regeneration of particulate traps for diesel exhaust cleaning. Chem. Eng. Sci. 58: 951 – 958.

[52] Neeft J.P.A., van Pruissen O.P., Makkee M., Moulijn J.A. (1997) Catalysts for the oxidation of soot from diesel exhaust gases II. Contact between soot and catalyst under practical conditions. Appl. Catal. B: Environmental 12: 21-31.

[53] Neyertz C.A., Miró E.E. Querini C.A. (2012) K/CeO2 catalysts supported on cordierite monoliths: Diesel soot combustion study. Chem. Eng. J. 181-182: 93-102.

[54] Gálvez M.E., Ascaso S., Tobías I., Moliner R., Lázaro M.J. (2012) Catalytic filters for the simultaneous removal of soot and NOx: Influence of the alumina precursor on monolith washcoating and catalytic activity. Catal. Today 191: 96-105.

[55] Gálvez M.E., Ascaso S., Moliner R., Jiménez R., García X., Gordon A., Lázaro M.J. (2011) Catalytic filters for the simultaneous removal of soot and NOx: Effect of CO2 and steam on the exhaust gas of diesel engines. Catal. Today 176: 134–138.

[56] Bortolozzi J.P., Banús E.D., Milt V.G., Gutierrez L.B., Ulla M.A. (2010) The significance of passivation treatments on AISI 314 foam pieces to be used as substrates for catalytic applications. Appl. Surf. Sci. 257: 495–502.

[57] Raj R.E., Parameswaran V., Daniel B.S.S. (2009) Comparison of quasi-static and dynamic compression behavior of closed-cell aluminum foam. Mat. Sci. Eng. A 526: 11–15.

[58] Kurtbas I., Celik N. (2009) Experimental investigation of forced and mixed convection heat transfer in a foam-filled horizontal rectangular channel Original. Int. J. Heat Mass Transfer 52: 1313–1325.

[59] Twigg M.V., Richardson J.T. (2002) Theory and applications of ceramic foam catalysts. Chem. Eng. Research. Design, 80: 183-189.

[60] Banús E.D., Milt V.G., Miró E.E., Ulla M.A. (2009) Structured catalyst for the catalytic combustion of soot: Co,Ba,K/ZrO$_2$ supported on Al$_2$O$_3$ foam. Appl. Catal. A: General 362: 129–138.

[61] Nor M.A.A.M., Hong L.C., Ahmad Z.A., Akil H.M. (2008) Preparation and characterization of ceramic foam produced via polymeric foam replication method. J. Mat. Process. Tech. 207: 235-239.

[62] Silva S.A., Brunelli D.D., Melo F.C.L., Thim G.P. (2009) Preparation of a reticulated ceramic using vegetal sponge as templating. Ceramic. Int. 35: 1575-1579.

[63] Mao X., Wang S., Shimai S., (2008) Porous ceramics with tri-modal pores prepared by foaming and starch consolidation. 34: 107-112.

[64] Yu J., Sun X., Li Q., Li, X. (2008) Preparation of Al$_2$O$_3$ and Al$_2$O$_3$–ZrO$_2$ ceramic foams with adjustable cell structure by centrifugal slip casting. Mat. Sci. Eng. A 476: 274-280.

[65] He X., Zhou X., Su B. (2009) 3D interconnective porous alumina ceramics via direct protein foaming. Mater. Lett. 65: 830-832.

[66] Barg S., Soltmann C., Andrade M., Kock D., Grathwohl G. (2008) Cellular Ceramics by Direct Foaming of Emulsified Ceramic Powder Suspensions. J. Am. Cerm. Soc. 91: 2823-2829.

[67] Ashby M.F., Evans A.G., Fleck N.A., Gibson L.J., Hutchinson J.W., Wadley H.N.G., Metal Foams: A Desing Guide. Butterworth-Heinemann. 2000. United States of America.

[68] Ciambelli P., Palma V., Russo P., Vaccaro S. (2002) Deep filtration and catalytic oxidation: an effective way for soot removal. Catal. Today 73: 363–370.

[69] Russo P., Ciambelli P., Palma V., Vaccaro (2003) Simultaneous Filtration and Catalytic Oxidation of Carbonaceous Particulates. Top. Catal 22: 123-129.

[70] Ciambelli P., Palma V., Russo P., Vaccaro S. (2002) Performances of a catalytic foam trap for soot abatement. Catal. Today 75: 471–478.

[71] Ciambelli P., Palma V., Russo P., Vaccaro S. (2005) Issues on soot removal from exhaust gases by means of radial flow ceramic traps. Chem. Eng. Sci. 60: 1619-1627.

[72] Palma V., P. Russo, M. D'Amore, P. Ciambelli (2004) Microwave regenerated catalytic foam: a more effective way for PM reduction. Top. Catal. 30/31: 261-264.

[73] Saracco G., Badini C., Russo N., Specchia V. (1999) Development of catalysts based on pyrovanadates for diesel soot combustion. Appl. Catal. B: Environmental 21: 233–242.

[74] Setiabudi A., van Setten B.A.A.L., Makkee M., Moulijn J.A. (2002) The influence of NO$_x$ on soot oxidation rate: molten salt versus platinum. Appl. Catal. B: Environmental 35: 159–166.

[75] Ambrogio M., Saracco G., Specchia V. (2001) Combining filtration and catalytic combustion in particulate traps for diesel exhaust treatment. Chem. Eng. Sci. 56: 1613-1621.

[76] Saracco G., Russo N., Ambrogio M., Badini C., Specchia V. (2000) Diesel particulate abatement via catalytic traps. Catal. Today 60: 33–41.

[77] Ciambelli P., Corbo P., Palma V., Russo P., Vaccaro S., Vaglieco B. (2001) Study of Catalytic Filters for Soot Particulate Removal from Exhaust Gases. Top. Catalysis 16-17: 279-284.

[78] Caglar B., Üner D. (2007) Preparation and Morphological Characterization of a Catalytic Soot Oxidation SiC Foam Filter. Turkish J. Chem. 31: 487-492.

[79] van Setten B.A.A.L., Bremmer J., Jelles S.J., Makkee M., Moulijn J.A. (1999) Ceramic foam as a potential molten salt oxidation catalyst support in the removal of soot from diesel exhaust gas. Catal. Today 53: 613–621.

[80] van Setten B.A.A.L., van Gulijk C., Makkee M., Moulijn J.A. (2001) Molten Salts Are Promising Catalysts. How to Apply in Practice? Top. Catal. 16/17: 275-278.

[81] Ciambelli P., Matarazzo G., Palma V., Russo P., Merlone Borla E., Pidria M.F. (2007) Reduction of soot pollution from automotive diesel engine by ceramic foam catalytic filter. Top. Catal. 42–43: 287-291.

[82] E. Bruneel, J. Van Brabant, M.T. Le, I. Van Driessche (2012) Deposition of a Cu/Mo/Ce catalyst for diesel soot oxidation on a sintered metal fiber filter with a CeO_2 anti corrosion coating, Catal Comm., in press.

[83] Walther G., Klöden B., Büttner T., Weißgärber T., Kieback B., Böhm A., Naumann D., Saberi S., Timberg L. (2008) A New Class of High Temperature and Corrosion Resistant Nickel-Based Open-Cell Foams. Adv. Eng. Mater. 10: 803-811.

[84] Kwon H.J., Kim Y., Nam I., Jung S.M., Lee J. (2010) The Hydrothermal Stability of Paper-Like Ceramic Fiber and Conventional Honeycomb-Type Cordierite Substrates Washcoated with Cu-MFI and V_2O_5/TiO_2 Catalysts for the Selective Reduction of NO_x by NH_3. Top. Catal. 53: 439-446.

[85] Bhardwaj N., Hoang V., Nguyen K.L. (2007) Effect of refining on pulp surface charge accessible to polydadmac and FTIR characteristic bands of high yield kraft fibres. Biores. Technol. 98: 962-966.

[86] Ichiura H., Kitaoka T., Tanaka H. (2003) Removal of indoor pollutants under UV irradiation by a composite TiO_2-zeolite sheet prepared using a papermaking technique. Chemosphere 50: 79-83.

[87] Ichiura H., Kitaoka T., Tanaka H. (2002) Preparation of composite TiO_2-zeolite sheets using a papermaking technique and their application to environmental improvement: Part I Removal of acetaldehyde with and without UV irradiation. J. Mater. Sci. 37: 2937-2941.

[88] Koga H., Ishihara H., Kitaoka T., Tomoda A., Suzuki R., H. Wariishi (2010) The Hydrothermal Stability of Paper-Like Ceramic Fiber and Conventional Honeycomb-Type Cordierite Substrates Washcoated with Cu-MFI and V_2O_5/TiO_2 Catalysts for the Selective Reduction of NO_x by NH_3. J. Mat. Sci. 45: 4151-4157.

[89] Koga H., Umemura Y., Ishihara H., Kitaoka T., Tomoda A., Suzuki R., Wariishi H. (2009) Paper-structured fiber composites impregnated with platinum nanoparticles synthesized on a carbon fiber matrix for catalytic reduction of nitrogen oxides. Appl. Catal. B: Environmental 90: 699-704.

[90] Fukahori S., Kitaoka T., Tomoda A., Suzuki R., Warrishi H. (2006) Methanol steam reforming over paper-like composites of Cu/ZnO catalyst and ceramic fiber. Appl. Catal. A: General 300: 155–161.

[91] Koga H., Fukahori S., Kiyaoka T., Tomoda A., Suzuki R., Wariishi H. (2006) Autothermal reforming of methanol using paper-like Cu/ZnO catalyst composites prepared by a papermaking technique. Appl. Catal. A: General 309: 263–269.

[92] Koga H., Ishihara H., Kitaoka T., Tomoda A., Suzuki R., Wariishi H. (2010) NOx reduction over paper-structured fiber composites impregnated with Pt/Al2O3 catalyst for exhaust gas purification. J. Mater. Sci. 45: 4151–4157.

[93] Koga H., Fukahori S., Kitaoka T., Nakamura M., Wariishi H. (2008) Paper-structured catalyst with porous fiber-network microstructure for autothermal hydrogen production. Chem. Eng. J. 139: 408–415.

[94] Fukahori S., Koga H., Kitaoka T., Tomoda A., Suzuki R., Wariishi H. (2006) Hydrogen production from methanol using a SiC fiber-containing paper composite impregnated with Cu/ZnO catalyst. Appl. Catal. A: General 310: 138–144.

[95] Krishna K., Makkee M. (2006) Soot oxidation over NOx storage catalysts: Activity and deactivation. Catal. Today 114: 48-56.

[96] Cecchini J.P., Serra R.M., Barrientos C.M., Ulla M.A., Galván M.V., Milt V.G. (2011) Ceramic papers containing Y zeolite for toluene removal. Micropor. Mesopor. Mat. 145: 51-58.

[97] López-Fonseca R., Landa I., Elizundia U., Gutiérrez-Ortiz M.A., González-Velasco J.R. (2007) A kinetic study of the combustion of porous synthetic soot. Chem. Eng. J. 129: 41-49.

[98] Nhon Y.N.H., Magan H.M., Petit C. (2004) Catalytic diesel particulate filter evaluation of parameters for laboratory studies. Appl. Catal. B: Environmental 49: 127–133.

[99] Peralta M.A., Gross M.S., Sánchez B.S., Querini C.A. (2009) Catalytic combustion of diesel soot: Experimental design for laboratory testing. Chem. Eng. J. 152: 234–241.

[100] Lizarraga L., Souentie S., Boreave A., George S.C., D'Anna B., Vernoux P. (2011) Effect of Diesel Oxidation Catalysts on the Diesel Particulate Filter Regeneration Process. Environ. Sci.Technol. 45: 10591–10597.

[101] Tanthapanichakoon W., Charinpanitkul T. (2012) Suppression of fugitive dust emitted from stone quarrying process using wetted wire screen. Sep. Purif. Technol. 92: 17-20.

[102] Caroca J.C., Millo F., Vezza D., Vlachos T., De Filippo A., Bensaid S., Russo N., Fino D. (2011) Detailed Investigation on Soot Particle Size Distribution during DPF Regeneration, using Standard and Bio-Diesel Fuels. Ind. Eng. Chem. Res. 50: 2650–2658.

[103] Bensaid S., Russo N. (2011) Low temperature DPF regeneration by delafossite catalysts. Catal. Today 176: 417– 423.

[104] Silva R.F., De Oliveira E., de Sousa Filho P.C., Neri C.R., Serra O.A. (2011) Diesel/biodiesel soot oxidation with CeO_2 and CeO_2-ZrO_2-modified cordierites: a facile way of accounting for their catalytic ability in fuel combustion processes. Quimica Nova, 34: 759-763.

[105] www.dieselnet.com/tech/engine_control.html.

Exhaust Gas After Treatment and EGR

Chapter 6

NO_x Storage and Reduction for Diesel Engine Exhaust Aftertreatment

Beñat Pereda-Ayo and Juan R. González-Velasco

Additional information is available at the end of the chapter

1. Introduction

Diesel and lean-burn engines provide better fuel economy and produce lower CO_2 emissions compared to conventional Otto gasoline engines. However, the NO_x gas components in the lean (oxidizing) exhausts from diesel and lean-burn engines cannot be efficiently removed with the classical three-way catalyst (TWC) under operating conditions with excess of oxygen in the exhaust gas. Among the available technologies under research, the NO_x storage-reduction (NSR) catalyst seems to be the most promising method to solve the problem. Basically, NSR catalysts consist of a cordierite monolith washcoated with porous alumina on which an alkali or alkali-earth oxide (e.g. BaO) and a noble metal (Pt) are deposited. These catalysts operate under cyclic conditions. During the lean period, when oxygen is in excess, the platinum oxidizes NO to a mixture of NO and NO_2 (NO_x), which is adsorbed (stored) on Ba as various NO_x species (nitrate, nitrite). During the subsequent short rich period, when some reductant (e.g. H_2) is injected, NO_x ad-species are released and reduced to nitrogen on Pt. Ammonia and N_2O byproduct formation upon NO_x reduction can also be observed over Pt-BaO/Al_2O_3 NSR catalysts.

In this chapter a systematic methodology for preparing Pt-Ba/Al_2O_3 NSR monolith catalysts is presented, the NO_x storage and reduction mechanisms on the catalyst are analysed, and the optimal control of different operational variables to achieve the NSR process with maximum production and selectivity to nitrogen is modeled [1].

2. Historical background

Main pollutants generated in the engine exhaust gases are nitrogen oxides (NO_x), carbon oxides (CO_x), hydrocarbons (HC) and particulate matter (PM). The last term is referred to small particles leaving the engine, mainly constituted by carbonaceous material. These fine particles can enter into the human lungs, being responsible for some breathing and cardiovascular diseases [2].

The hydrocarbons are organic volatile compounds able to form ozone smog at the ground level when interacting with nitrogen oxides under the sun light. Ozone irritates the eyes, hurts the lungs, causes asthma attack and aggravates other respiratory problems. In addition, ozone is one of the primary components of photochemical smog (or just smog for short). Furthermore, hydrocarbons can also cause cancer [3].

Nitrogen oxides, same as hydrocarbons, are precursors for ozone formation. The NO_2 contributes importantly to the formation of acid rain [4]. The carbon monoxide (CO) reduces the oxygen flow in the blood and results particularly dangerous for people with heart diseases [5]. The carbon dioxide (CO_2) is a greenhouse gas able to make an atmosphere layer trapping the heat and contributing to the global warm of the earth [6].

2.1. Legislation

The negative impacts of those emissions on the human health and the environment and climate have forced legislation to control and limit such emissions. In the U.S.A., NO_x emissions from mobile sources contribute almost 50% of those produced in total, so that more and stricter regulations have been introduced for reducing NO_x emissions from the automobiles [7].

Table 1 shows the most important regulations as introduced by the European Union from the first directive Euro 1 (1992), then Euro 2 (1996), Euro 3 (2000), Euro 4 (2005), Euro 5 (2009), and the most recent Euro 6 (2014) [8]. Emission limits for CO, HC, NO_x, and PM were proposed for petrol and diesel engines. Former regulations limited HC+NO_x jointly, which later were split up into individual HC and NO_x limits.

Step	Date	CO	HC	HC+NO_x	NO_x	PM
		Diesel engines				
Euro 1	07.1992	2.72	—	0.97	—	0.14
Euro 2	01.1996	1.00	—	0.70	—	0.08
Euro 3	01.2000	0.64	—	0.56	0.50	0.05
Euro 4	01.2005	0.50	—	0.30	0.25	0.025
Euro 5	09.2009	0.50	—	0.23	0.18	0.005
Euro 6	09.2014	0.50	—	0.17	0.08	0.005
		Petrol engines				
Euro 1	07.1992	2.72	—	0.97	—	—
Euro 2	01.1996	2.20	—	0.50	—	—
Euro 3	01.2000	2.30	0.20	—	0.15	—
Euro 4	01.2005	1.00	0.10	—	0.08	—
Euro 5	09.2009	1.00	0.10	—	0.06	0.005
Euro 6	09.2014	1.00	0.10	—	0.06	0.005

Table 1. EU emission standards for passenger cars, g km^{-1}.

On the other hand, Euro standards have been completed with stricter regulations for sulphur content in fuels. In fact, the content of S in diesel could not surpass 350 ppm from the year 2000, and only 50 ppm from 2005 (for petrol, 150 ppm in 2000 and 50 ppm in 2005). From 2009, S-free fuels (S ≤ 10 ppm) have been implemented.

In the most recent Euro standards the durability of the catalyst is also specified, e.g. Euro 3 required the emission standards for 80,000 km or 5 years (whatever first occurs). Following regulations required 100,000 km or 5 years. From 2000, with the entrance of Euro 3, vehicles should be equipped with on board diagnostics (OBD), announcing to the driver the system damage or wrong operation, then causing higher emissions which should be avoided.

2.2. Automobile exhaust aftertreatment

More exigent legislation on automobile exhaust emissions has led to the development of aftertreatment systems. Today, three way catalysts (TWC) oxidize CO and HC to CO_2 and H_2O, and simultaneously reduce NO_x to N_2 in a very efficient way for conventional Otto gasoline engines [9-12]. The conventional gasoline engines operate at stoichiometric air/fuel ratio, A/F=14.63 (w/w) [13,14], which produces an exhaust gas with the exact balance of CO, H_2 and HC (reducing species) needed to reduce NO_x and O_2 (oxidizing species). However, diesel engines operate with higher A/F ratios, from 20:1 to 65:1 [15], then producing an exhaust gas with oxygen in excess (Table 2 [16]).

		Conventional gasoline engine	Diesel engine	Lean engine
O_2	%vol.	0.2 – 2	5 – 15	4 – 18
CO_2	%vol.	10 – 13.5	2 – 12	2 – 12
H_2O	%vol.	10 – 12	2 – 10	2 – 12
N_2	%vol.	70 – 75	70 – 75	70 – 75
CO	%vol.	0.1 – 6	0.01 – 0.1	0.04 – 0.08
HC	%vol. C_1	0.5 – 6	0.005 – 0.05	0.002 – 0.015
NO_x	%vol.	0.04 – 0.4	0.003 – 0.06	0.01 – 0.05
SO_x	Related to the content of S in the fuel			

Table 2. Exhaust gas composition, depending on the type of engine.

Fig. 1 shows the conversion curves for each pollutant as a function of the air/fuel ratio, for a TWC. Around the stoichiometric point (A/F=14.63), all the three pollutants (HC, CO and NO) are highly converted (>95%), i.e. they are almost totally removed. However, when the environment is abundant in oxygen as in diesel engines (A/F>20), although this environment enhances the oxidation of HC and CO, the reduction of NO becomes practically inefficient, then this pollutant cannot be appropriately removed with TWC technology [15,17,18].

On the other hand, technical solutions existing for the optimal compromise in removal of NO_x/PM [19], by exhaust gas recirculation (EGR), are not able to achieve the requirements of Euro 6. In fact, in these systems, reduction of PM means eventually an increment of NO_x and viceversa [20]. Consequently, current technologies combining diesel particulate filters (DPF) and DeNO$_x$ catalysts [21-23] are being reconsidered.

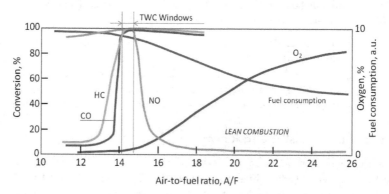

Figure 1. Fuel consumption and TWC behaviour of stoichiometric petrol engines, related to the air-to-fuel ratio.

At present, the removal of NO$_x$ in the diesel engine exhaust gases, mainly in heavy-duty lorries, is controlled by selective catalytic reduction (SCR) with ammonia generated by hydrolysis of urea which must be stored in an on-board container [24,25]. For light vehicles and passenger cars running under lean conditions, the NH$_3$-SCR technology is not appropiate because of the volume of the needed ammonia container. Thus, other technologies are being developped, including the SCR with the presence of reductants in the exhaust, e.g. hydrocarbon [26], and the NO$_x$ storage and reduction (NSR) [27-34], which seems to be the most promising technology and to which is dedicated in this chapter.

2.3. General aspects of the NSR (NO$_x$ Storage and Reduction) catalysis

Up to day, the NSR is considered as the most promising technology for NO$_x$ removal from diesel engine exhaust gases. The corresponding devices are also denominated lean NO$_x$ traps (LNT). Recent excellent revisions can be found in the literature on this technology [34-36]. Following is a brief summary of the chemical principles used in NSR as to facilitate understanding of next sections.

The NSR catalysts run cyclically under lean environment (oxidizing) and rich environment (reducing), being defined by the corresponding A/F ratios. The concept was introduced by Toyota in the middle 90s [27,33]. While running on the road, lean and rich conditions have to be used in an alternative way [37,38]. Under lean conditions, with excess of oxygen (high A/F), NO$_x$ are adsorbed on the catalyst, and then under rich conditions (A/F<14.63) the stored NO$_x$ are released and reduced. Consequently, an NSR catalyst needs sites for NO$_x$ adsorption (alkaline or earth-alkaline compounds) and also sites for NO$_x$ oxidation and/or reduction (noble metals, as in the TWC technology). Most studies in the literature have used storage materials based on Ba. Also other metals such as Na, K, Mg, Sr and Ca have been used. Thermodynamic and kinetic data demonstrated that basicity of alkaline and earth-alkaline metals is related directly to the NOₓ storage capacity, i.e. the storage behavior at 350 °C decreases as follows: K > Ba > Sr ≥ Na > Ca > Li ≥ Mg [34].

The noble metals are normally incorporated with very low percentage, 1-2 wt%. As in the TWC technology, platinum, palladium and rhodium are mostly used [39]. The metal participates into two important steps of the NSR mechanism, the oxidation of NO to NO₂ during the lean period and the reduction of NOₓ released during the rich period. In general, it is stablished in the literature that Pt is a good catalyst for NO oxidation, while Rh is more active for NOₓ reduction. Obviously, the storage compounds as well as the noble metals should be dispersed on porous materials with high surface area (Al₂O₃, ZrO₂, CeO₂, MgO) washcoated over a monolithic structure, usually cordierite. The most studied formulation in the literature has been Pt-Ba/Al₂O₃, which has also been chosen for this study.

Presently, it is well assumed that the NSR mechanism can be explained by the five following steps, as represented in the upper scheme of Fig. 2 [34,35]:

a. Oxidation of NO to NO₂ (lean conditions, oxidizing environment).
b. Adsorption of NOₓ as nitrites or nitrates on the storage sites (lean period, oxidizing environment).
c. Injection and evolution of the used reductant agent (H₂, CO or HC).
d. Release of the stored NOₓ from the catalyst surface to the gas stream (rich period, reducing environment).
e. Reduction of NOₓ to N₂ (rich period, reducing environment).

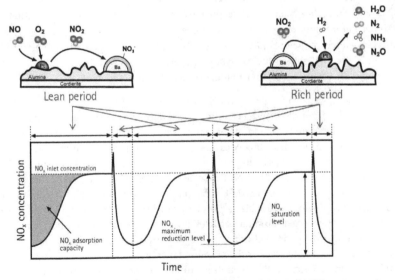

Figure 2. Storage and reduction of NOₓ. (a) Schematics of the mechanism; (b) NOₓ concentration curves at the exit, during lean and rich periods.

The typical NO_x storage and reduction behaviour can be observed in the bottom graph of Fig. 2. At the beginning of the lean period nearly all the NO_x ($NO+NO_2$) entering the trap is adsorbed, afterwards the NO_x outlet concentration progressively increases due to the successive saturation of the available trapping sites. When saturation is completed, NO_x outlet concentration equals the NO_x inlet concentration. During the subsequent rich period, when H_2 is injected, the adsorbed NO_x species on the catalyst surface react with hydrogen to form N_2O, NH_3 or N_2, resulting in the regeneration of the trap which is again ready for the following lean period.

3. Preparation procedure of monolithic NSR catalysts

Most work dealing with NO_x storage and reduction technology have normally used powder catalyst to carry out different studies. However, for real application, NSR catalysts have to be synthesized in a monolithic structure in order to minimize the pressure drop in the catalytic converter [40-42]. The preparation procedure of powder or monolithic catalysts differs notably. While conventional techniques are used for the incorporation of the active phases in powder catalysts, such as wetness impregnation [43-45], the synthesis of monolithic catalysts requires more sophisticated techniques. This section will be focused on the preparation procedure of monolithic NSR catalysts, paying special attention on their final physico-chemical characteristics (dispersion and distribution of the active phases) and their correlation with the activity for NO_x storage and reduction.

In real application, the mechanical properties of the catalyst are crucial due to the dramatic temperature changes and vibrational strengths that are expected. In this sense, cordierite ($2MgO.2Al_2O_3.5SiO_2$) has been chosen as the base material in automotive application due to its high thermal stability and low expansion coefficient. However, this material exhibits a low surface area which is not suitable for the subsequent incorporation of the active phases. Consequently, the first step of the catalyst preparation consists on the monolithic substrate washcoating with a high surface area oxide, usually alumina.

The most common washcoating procedure is carried out by dipping the monolith into slurry which contains the alumina for washcoating. The monolith is immersed in the slurry for a few seconds and then removed, and the excess of liquid remaining in the channels is blown out with compressed air. This procedure is repeated until the desired Al_2O_3 weight is incorporated as washcoat. It has been previously reported that the characteristics of the final coated monoliths are governed by the properties of the slurry [42], considering as main variables the Al_2O_3 particle size, the Al_2O_3 wt% in the slurry and the pH. Agrafiotis et al. [46] found a threshold value of particle size around 5 μm which is coincident with the size of the cordierite macropores; larger alumina particles do not penetrate into the macropores of the substrate resulting in a poor anchoring of the alumina layer. Therefore, the smaller the particle size in the slurry, the higher the alumina layer anchoring is. In fact, in our previous work [47], the immersion of the monoliths in an alumina slurry with a particle size distribution centered in 1 μm, led to a highly adhered alumina layer, with a weight loss smaller than 0.25% after the washcoated monolith was immersed in ultrasound bath for 15 min. Regarding the Al_2O_3 wt% in the slurry, two contradictory effects are observed. On one hand, as the slurry concentrates in Al_2O_3, few immersions are required to achieve a given amount of washcoated alumina, but on the other hand, the increase in the slurry viscosity resulted in non-homogeneous coating. The influence of the slurry viscosity on the alumina layer homogeneity has been associated with the ease for the suspension excess to be blown out from the monolith channels [48-50]. Another characteristic to be controlled is the stabilization of the alumina slurry so as to avoid the particles from settling down. Nijhuis et al. [41] suggested that addition of some acid to shift pH between 3 and 4 improved the slurry stabilization. Furthermore, the addition of acetic acid up to 2.5 mol l⁻¹ (pH=2.6) decreased considerably the viscosity of the slurry, permitting the use of concentrated Al_2O_3 slurries without penalization in the layer homogeneity [47].

Fig. 3 shows the characterization of a washcoated monolith with scanning electron microscopy (SEM). The washcoating procedure was carried out by immersion of the monolith into alumina slurry with the following characteristics: 20 wt% Al_2O_3, mean diameter of 1 μm and 2.5 mol l⁻¹ of acetic acid. Eight immersions of the monolith were needed to achieve around 400 mg of Al_2O_3 over the monolithic substrate (D=L=2 cm). Fig. 3a shows a lower magnification image where the intersection of different channels of the monolith can be observed. As it can be clearly noticed, the original structure of the cordierite was completely covered with alumina. The deposition is preferential in the corners of the channels whereas far away from this position, the alumina layer has a constant thickness of 5 μm. Fig. 3b shows a higher magnification image where a crack in the alumina layer is observed. This image also confirms that the alumina layer was composed of particles around 1 μm in size.

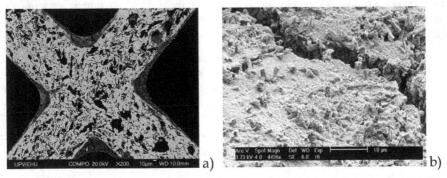

Figure 3. Scanning electron microscoy (SEM) images of the washcoated monolith. (a) Cross section. (b) Higher magnification image of surface in the monolith corner.

The next step in the catalyst preparation is the incorporation of the active phases. As already mentioned, NSR catalysts are usually composed of an alkali or alkali-earth oxide and a noble metal deposited onto the alumina. The most common metal used for NSR catalyst formulation is Pt, whereas BaO is normally used as the storage component [34-36]. The order of the incorporation steps of the active phases Pt and Ba is crucial, especially when operating at higher temperatures; a higher storage capacity is obtained when impregnating Pt/Al_2O_3 with Ba than when impregnating Ba/Al_2O_3 with Pt, increasing the storage value as much as 54% when adding Ba in the last step [51].

Platinum was incorporated following two different procedures, conventional wetness impregnation (1 monolith) and adsorption from solution (3 monoliths). For the conventional procedure, the channels of the monolith were filled with an aqueous solution containing the desired amount of Pt, using $Pt(NH_3)_4(NO_3)_2$ as a precursor [52]. Then, liquid was evaporated at 80 °C and finally the monolith was calcined at 500 °C for 4 h. On the other hand, in the adsorption from solution procedure, the monoliths were immersed in an aqueous solution with the adequate concentration of Pt. The pH of the solution was turned basic (11.9) in order to generate an electrostatic attraction between the alumina surface, positively charged, and the Pt precursor, negatively charged $Pt(NH_3)_4^{2+}$ [53-55]. The monoliths were maintained immersed in the solution for 24 h so as to reach the adsorption equilibrium. Then, the monoliths were removed from the solution, the excess of liquid blown out and finally the monoliths were calcined at 450, 500 and 550 °C, respectively. The four prepared monolith catalysts were tested for their performance in the NSR process.

Irrespective of the calcination temperature, the monolith prepared by wetness impregnation showed the lowest dispersion of platinum (15%). In the rest of the samples, platinum dispersion decreased as the calcination temperature increased, from 54% at 450 °C to 46% at 500 °C and finally to 19% at 550 °C. Then, 500 °C was chosen as the optimal calcination temperature as a good compromise between platinum dispersion and thermal stabilization of the catalyst. Fig. 4b shows the platinum particle size distribution determined from the transmission electron microscopy image (Fig. 4a) for a Pt/Al_2O_3 sample prepared by adsorption from solution and calcined at 500 °C. As it can be observed, the Pt particles are fairly dispersed over the alumina washcoat with a mean particle size of 1.3 nm.

The last step in the NSR catalyst preparation is the incorporation of the NO_x storage component, i.e. barium. The precursor used was barium acetate [52] and two different procedures were followed: wetness impregnation and incipient wetness impregnation (also known as dry impregnation). For wetness impregnation, the monolith channels were filled with an aqueous solution containing the desired amount of barium. Then, the monolith was dried and calcined. Alternatively, for incipient wetness impregnation, the monolith was immersed in an aqueous solution with an adequate concentration of barium acetate for a few seconds; then, the monolith was removed and the liquid in the channels was blown out with compressed air. Thus only the liquid retained in the pores of the alumina remained in the monolith. In order to determine the distribution of Ba in the catalyst, the monolith was divided into 8 pieces and the content of barium was determined by inductively coupled plasma mass spectrosmetry (ICP-MS). It was found that the distribution of barium resulted in an egg-shell type for wetness impregnation, whereas the incorporation of Ba by incipient wetness impregnation led to more homogenous distribution. Table 3 resumes the preparation procedure, the catalyst physico-chemical characteristics and the NO_x storage achieved with the prepared catalysts (A, B, C and D).

a)

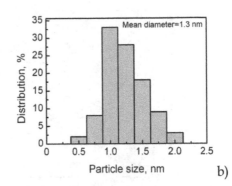
b)

Figure 4. Platinum dispersion measurements. a) Transmision electron microscopy image. b) Platinum particle size distribution.

The activity of the prepared catalysts was tested in a vertical downstream reactor with a feedstream composed of 380 ppm NO and 6% O_2 during the lean period (150 s) and 380 ppm NO and 2.3% H_2 during the rich period (20 s) using nitrogen as the balance gas in both cases. The total flowrate was 3365 ml min^{-1} that corresponds to a gas hourly space velocity (GHSV) of 32,100 h^{-1}. Fig. 5 shows the NOₓ concentration profile at the reactor exit for A, B, C and D catalysts. As it can be observed, the NOₓ concentration profile is always below the inlet value (380 ppm NO) which evidences the activity of the prepared catalysts for the storage of NOₓ. During the storage-reduction cycles, the typical NOₓ concentration profile was recorded [56,57]. At the beginning of the lean period practically all the NOₓ is stored, and consequently its concentration at the reactor exit is very low. Then, as the lean period time increases, the storage sites become saturated and the NOₓ concentration at the reactor exit gradually increases. During the rich period, the NOₓ stored are released and reduced with the injected hydrogen, leaving the catalyst surface clean for the subsequent storage period.

	A	B	C	D
Pt incorporation	WI	ADS	ADS	ADS
Calcination T, °C	500	550	500	500
Dispersion,%	15	19	46	46
*Distribution	+	+++	+++	+++
% Pt	0.72	1.43	1.34	1.14
Ba incorporation	WI	WI	WI	DI
*Distribution	+	+	+	++
% BaO	13.1	13.3	16.3	25.2
NOₓ storage capacity, %	47.5	55.7	69.6	76.7

Table 3. Preparation procedure, physico-chemical characteristics and NOₓ storage capacity for the prepared catalysts A, B, C and D. WI: Wetness impregnation; ADS.: Adsorption from solution; DI: Dry impregnation. The dispersion values are estimated based on powder Pt/Al_2O_3 samples. * Distribution: (+) not good (++) good (+++) very good.

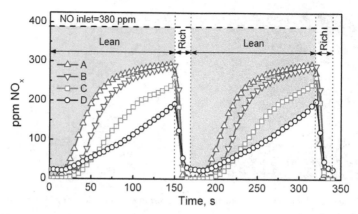

Figure 5. NOₓ concentration profile at the reactor exit for two consecutive NOₓ storage and reduction cycles for catalysts A, B, C and D.

The NOₓ storage capacity is related with the area between the NO inlet level and the NOₓ outlet concentration profile; the lower NOₓ concentration at the reactor outlet the higher activity of the catalyst for the storage of NOₓ. Thus, among the prepared catalysts, catalyst D was found to be the most active (gray area on the rigth graph) and catalyst A (blue area on the left graph). Quantification of the NOₓ storage capacity can be found in Table 3 with the following order from the least to the most active: A<B<C<D, according to the physico-chemical characteristics of the samples. It is well known that the Pt-Ba pair is the responsible for the storage of NOₓ and that proximity between both metals is beneficial for the process [47,58,59]. Consequently, catalyst A resulted in the less active sample due to the low platinum dispersion and non-homogeneous distribution of both Pt and Ba, as it was prepared by conventional wetness impregnation (WI) of Pt and Ba. For catalyst B, the incorporation of Pt by adsorption from solution (ADS) increased Pt dispersion, but just slightly as the higher calcination temperature (550 °C) also provokes some platinum sintering. Higher Pt loading and dispersion were identified as responsible for better storage capacity, from catalyst A (47.5%) to catalyst B (55.7%). The lower calcination temperature (500 °C) for catalyst C provided much higher dispersion, thus enhancing the NOₓ storage capacity up to 69.6%. Furthermore, the incorporation of Ba by dry impregnation (DI) in catalyst D provided a better distribution of barium over the monolith and consequently increased the Pt-Ba proximity resulting in the best storage capacity (76.7%), i.e. 76.7% of NO at the inlet was trapped in the catalyst.

4. The chemistry of NOₓ storage and reduction

Many studies are available in the technical literature dealing with the storage step of NSR catalysts. Particularly relevant in this field is research by Forzatti et al. [60-65] and Fridell et al. [66-68]. In situ FTIR spectroscopy has been found to be a very useful tool and several studies have been made on adsorbed NOₓ species, though the assignment of peaks is still under debate. Takahashi et al. [27] were pioneers in studying the interaction of NOₓ over

NSR catalysts and assigned the 1350 cm^{-1} peak to the nitrate anion. Fig. 6a shows the FTIR spectra of Pt-Ba/Al$_2$O$_3$ sample after it was exposed during 20 minutes to a feedstream composed of 440 ppm NO and 7% O$_2$ using N$_2$ as the balance gas. As it can be observed, the FTIR spectra changed very significantly with the operating temperature. Bridged nitrites situated at 1220 cm^{-1} [66] were dominant when the adsorption was carried out below 250 °C, whereas the dominant species became ionic nitrates (asymmetric and symmetric modes of monodentate nitrates) located at 1332 and 1414 cm^{-1} [69] for temperatures above 250 °C. This shift in the adsorption mode from nitrites to nitrates with temperature had been already reported in the literature [62,63,66,67,70].

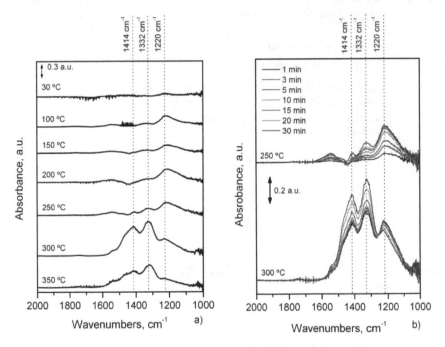

Figure 6. FTIR experiments with powder Pt-BaO/Al$_2$O$_3$. (a) Absorbance signals at different temperatures and after 20 min of contact time. (b) Absorbance of the sample at 250 and 300 °C with increasing contact time.

All spectra included in Fig. 6a were recorded after the sample had been exposed to the lean gas mixture for 20 min. However, it can be interesting to examine the evolution of adsorbed species with increasing contact time. Owing to the fact that the adsorption mode of NO$_x$ changed from nitrites at 250 °C to nitrates at 300 °C, several FTIR spectra were recorded at different contact times for those temperatures. The first spectrum was recorded after the sample had been exposed to the lean gas mixture for 1 min while the last one was taken after 30 min. As revealed by Fig. 6b, when the adsorption was carried out at 250 °C nitrite was immediately formed upon admission of NO, whereas nitrate formation was delayed.

Furthermore, the intensity of the peaks corresponding to nitrite (1220 cm^{-1}) and nitrate (1322, 1414 cm^{-1}) increased nearly in the same extent with increasing contact time, which means that there was no conversion from nitrites to nitrates or that conversion from nitrites to nitrates and formation of additional nitrite species occurred simultaneously. In short, below 250 °C nitrite was the dominant adsorption species even at contact times of 30 min, which is much longer than in real operation (1-2 min).

On the other hand, the adsorption pattern resulted completely different at 300 °C (Fig. 6b). From the beginning of the adsorption, the intensity of the bands assigned to ionic nitrates was higher than nitrite. Moreover, it can be noticed that the adsorption peak assigned to nitrites resulted maximum in the first minute of storage and then gradually decreased till minimum after 30 min of contact time. Thus, it can be concluded that there is a shift from nitrite to nitrate when increasing contact time which can be associated with the oxidation of nitrites to nitrates under the lean gas mixture.

In early ages of NSR catalysts, Fridell et al. [66] proposed a three step mechanism in which NO_2 is at first loosely adsorbed on BaO as a BaO-NO_2 species; this species then decomposes to BaO_2 and NO (which is released in the gas phase) and finally barium peroxide reacts with the gas-phase NO_2 to give barium nitrate which can be illustrated as:

$$NO_2 + BaO \rightarrow BaO\text{-}NO_2 \tag{1}$$

$$BaO\text{-}NO_2 \rightarrow BaO_2 + NO \tag{2}$$

$$2NO_2 + BaO_2 \rightarrow Ba(NO_3)_2 \tag{3}$$

The overall stoichiometry of NO_2 adsorption implies the release of one molecule of NO for the consumption of three molecules of NO_2. This reaction is known as the NO_2 disproportionation and has been widely reported for NSR catalysts [62,71-74]:

$$3NO_2 + BaO \rightarrow Ba(NO_3)_2 + NO \tag{4}$$

The formation of nitrate species following the reactions above described, clearly evidences that the oxidation of NO to NO_2 is a preliminar and necessary step for the adsorption of NO. The reaction mechanism used to describe the NO oxidation consists of the following adsorption and desorption steps [75,76]:

$$O_2 + 2Pt \rightarrow 2Pt\text{-}O \tag{5}$$

$$NO + Pt \rightarrow NO\text{-}Pt \tag{6}$$

$$NO + Pt\text{-}O \rightarrow NO_2\text{-}Pt \tag{7}$$

$$NO\text{-}Pt + Pt\text{-}O \rightarrow NO_2\text{-}Pt + Pt \tag{8}$$

$$NO_2\text{-Pt} \rightarrow NO_2 + Pt \tag{9}$$

On the other hand, FTIR experiments showed that apart from nitrates, surface nitrites are also formed during adsorption of NO over the Pt-Ba/Al$_2$O$_3$ catalyst. It has been proposed that barium peroxide formed in reaction (2) could also react with NO to form nitrites [62,71,77]:

$$BaO_2 + 2NO \rightarrow Ba\left(NO_2\right)_2 \tag{10}$$

BaO$_2$ can also be formed by an alternative route to reactions (1) and (2), in which NO$_2$ is not involved, as the following

$$O\text{-Pt} + BaO \rightarrow BaO_2 + Pt \tag{11}$$

The close proximity of BaO to Pt sites promotes spillover of the oxygen adatoms from Pt to BaO. From FTIR spectra shown in Fig. 6a it can be deduced that, at 300 °C, Pt catalyzes the formation of barium nitrate species from nitrite species, which is illustrated as:

$$Ba\left(NO_2\right)_2 + 2O\text{-Pt} \rightarrow Ba\left(NO_3\right)_2 + 2Pt \tag{12}$$

Thus, from the experiments shown in Fig. 6, two parallel routes can be described for the adsorption of NO$_x$, which are in concordance with the most accepted mechanism by Forzatti et al. [65]. The first route is called "nitrite route" where NO is oxidized at Pt sites and directly stored onto Ba neighbouring sites in the form of nitrite ad-species (reaction 10), which can be progressively transformed into nitrates depending on reaction temperature (reaction 12). The second route is called "nitrate route" which implies the oxidation of NO to NO$_2$ on Pt sites, followed by NO$_2$ desproportionation on Ba sites to form nitrates with the giving off NO into the gas phase (reaction 4).

The regeneration step of NSR catalysts is not so well understood as the storage step. Several studies have been published on the chemistry and mechanisms that rule the reduction of NO$_x$ ad-species by H$_2$. The nitrite and nitrate decomposition can be driven by either the heat generated from the reducing switch [78,79], or the decrease in oxygen concentration that lowers the equilibrium stability of nitrates [34,80]. However, under near isothermal conditions, it has been found that the reduction process is not initiated by the thermal decomposition of the stored nitrates, but rather by a catalytic pathway involving Pt [45]. The reduction of stored nitrates and nitrites leads to the formation of different nitrogen containing species, such as N$_2$, NH$_3$ and N$_2$O along with H$_2$O. The objective of the NSR operation is to maximize the conversion of NO into N$_2$, avoiding the formation of NH$_3$ and N$_2$O as far as possible. The operational conditions to run efficiently the NSR process are discussed in detail in section 5.

Fig. 7 shows the concentration profiles of NO, NO$_2$, NH$_3$, N$_2$O and H$_2$O, and the evolution of the MS-signal for N$_2$, O$_2$ and H$_2$ during the regeneration step when the reaction was carried out at 330 °C. The feedstream composition during the lean period was 975 ppm NO, 6% O$_2$ and Ar to balance, extending the length of this period until complete saturation of the catalyst was obtained. Afterwards, during the rich period, oxygen was replaced by 0.6% H_2 for 500 s. As can be observed in Fig. 7, before the regeneration period started the sum of NO and NO_2 concentration was close to the inlet value (975 ppm), confirming that the catalyst was saturated. The presence of NO_2 at the reactor exit is due to the oxidation of NO by Pt sites as described in eqns. (5)-(9). When the rich feedstream contacts the catalyst (t=0), the NO and NO_2 concentrations are progressively reduced, eventually reaching 0 ppm. At the very beginning of the rich period a sudden increase in the NO and NO_2 concentrations can be observed due to the release of adsorbed NO_x as a consequence of the decrease in oxygen partial pressure that reduces the stability of the stored nitrates and nitrites. Meanwhile, the incoming H_2 reacts with adsorbed NO_x to form N_2, NH_3 and N_2O. As can be observed in Fig. 7, the formation of N_2 and N_2O is detected immediately after the reduction period started, whereas the detection of H_2O and NH_3 was delayed, the later in a much more extent. On the other hand, the complete consumption of H_2 during the initial period of the regeneration together with the rectangular shape of the H_2O and N_2 formation curve indicates a "plug-flow" type of the regeneration mechanism. As several authors have already reported [45,78,81,82], the hydrogen front travels through the catalyst bed with complete regeneration of the trapping sites as it propagates down the bed with regeneration time. After the required time, complete regeneration of the trap is obtained, i.e. no nitrates or nitrites are present in the catalyst surface, and consequently H_2 is detected at the reactor outlet.

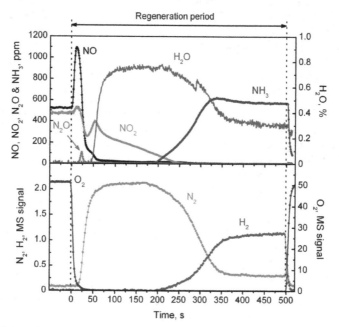

Figure 7. Evolution of NO, NO_2, N_2O, NH_3 and H_2O concentrations by FTIR and MS signals of O_2, N_2 and H_2, during Pt-BaO/Al_2O_3 catalyst regeneration at 330 °C.

The reactants and product profiles shown in Fig. 7 are in agreement with mechanistic aspects of the regeneration already reported [45,81,83,84]. The reduction of stored nitrates with hydrogen has been found to occur by the following reactions:

$$Ba(NO_3)_2 + 8H_2 \rightarrow 2NH_3 + BaO + 5H_2O \qquad (13)$$

$$Ba(NO_3)_2 + 5H_2 \rightarrow N_2 + BaO + 5H_2O \qquad (14)$$

Lietti et al. [84] reported that during reduction of stored nitrates at 100 °C, reaction (13) accounted for almost all the H_2 consumption, demonstrating that stored nitrates were reduced efficiently and selectively (>90%) to ammonia. On increasing the reduction temperature, nitrogen formation was promoted due to reaction (15) where the formed ammonia continued to react further with stored nitrates to form nitrogen.

$$Ba(NO_3)_2 + 10NH_3 \rightarrow 8N_2 + 3BaO + 15H_2O \qquad (15)$$

Thus, nitrogen formation involves a two-step pathway: the fast formation of ammonia by reaction of nitrates with H_2 (reaction 13) and the subsequent conversion of the ammonia formed with stored nitrates leading to the selective formation of N_2 (reaction 15). This overall mechanism for nitrates reduction during LNT regeneration has been confirmed by Pereda-Ayo [85] using isotope labelling techniques and explains the evolution of products at the reactor exit shown in Fig. 7. When the hydrogen front enters the catalyst, the stored NOₓ are thought to be converted mainly to ammonia, with total H_2 consumption. Then, the ammonia formed in the regeneration front reacts further with stored nitrates located downstream to give nitrogen. Thus, N_2 is detected as soon as the regeneration period starts, but no ammonia can be detected since it was completely consumed. As the regeneration time increases and the hydrogen front moves forward, the ammonia formed has fewer nitrates to react with, and therefore some NH₃ starts to leave the catalyst unreacted, being detected at the reactor outlet.

5. Analysis of engineering variables of the NSR process

The importance of the catalyst properties, chemical composition, structure, morphology and, in special dispersion and distribution of the metallic phases on its behavior in storing and reducing NOₓ has been well reviewed in the work of Roy and Baiker [35]. Generally, research in the scientific open literature has studied independently the two stages, storage and reduction, to advance in the understanding of the mechanisms that happen in each stage. Up to now many papers on the stage of storage have been published [e.g. 60,63,66-68,70,72,87], but notably less on the stage of regeneration (liberation and reduction) [e.g. 39,81,86,88,89]. However, very few studies have considered the whole operation, where the lean and rich periods occur successively as in the real application.

In fact, still scarce relations have been proposed between storage, regeneration and product distribution as determined in laboratory, and the optimization of the conditions in which the catalytic converter should operate in automobiles. It must be mentioned that engineering parameters, such as gas hourly spatial velocity (GHSV), residence time in the converter, and the lasting time of the periods of storage and regeneration, influence significantly the behavior of lean NO$_x$ traps (LNTs). Kabin et al. [90] studied the storage and reduction of NO$_x$ on model monolithic Pt-BaO/Al$_2$O$_3$ catalysts to relate the percentage of the NO$_x$ trapped during storage (trapping efficiency) and the reduced NO$_x$ percentage (average NO$_x$ conversion) with the load of Ba (6-25%) and the GHSV (30,000-120,000 h^{-1}). These authors concluded that the dependence of trapping efficiency on the storage period duration provides a good estimation of the time needed to get a given average conversion during the whole NSR process. More recently, Clayton et al. [91] determined the effects of the catalyst temperature, the composition of the rich stream (NO, H$_2$, O$_2$), duration of lean and rich periods and the H$_2$/NO ratio on the average conversion and product selectivity of a commercial Pt-BaO/Al$_2$O$_3$ catalyst. The NO$_x$ average conversion resulted maximum at 300 °C, and also the trapping efficiency resulted maximum at the same temperature. The selectivity to N$_2$ exhibited a maximum at slightly superior temperature, where the selectivity to NH$_3$ was minimized.

Thus, to define the optimal operation in the NSR technology, the efficiency of both the storage and reduction steps but also the global NSR efficiency must be studied. This section is devoted to set a definition for the global NSR efficiency of the process, obviously related to the catalyst behavior, the NO$_x$ storage and reduction mechanisms, and the kinetics of the release and reduction of NO$_x$ during the regeneration phase. The defined parameter must account the byproduct formation to maximize the NO$_x$ reduction efficiency towards N$_2$. The process selectivity depends notably on the lean and rich period duration. For this purpose, experimental runs have been carried out with storage period duration in the order of minutes, followed by rich injection periods during some seconds. The effect of the lean and rich period duration and the reductant concentration should be analyzed, stating as objective functions for optimization the storage capacity in the lean period, the NO$_x$ conversion in the rich period, and the selectivity towards N$_2$O/NH$_3$/N$_2$. Also a bidimensional analysis of operational variables, including temperature and hydrogen concentration, will be made in this section.

The experiments were carried out with a homemade 1.2%Pt-15%BaO/Al$_2$O$_3$ monolith catalyst, prepared as explained in section 3 [1,47]. Each time some operational variable was altered, at least ten successive lean-rich cycles were proceeded in order to assure a new stable state of the system, and then the performance was monitored.

5.1. Definition of response parameters

To evaluate the performance of the catalyst during the lean and rich periods the NO$_x$ storage capacity, NO$_x$ conversion and N$_2$/NH$_3$ selectivities have been determined from the concentration curves at the reactor exit monitored during the experimental storage-reduction cycles.

The total NO$_x$ stored during the lean period, was calculated as

$$NO_x^{stored}(\mu mol\ NO)=(NO^{in})_L - (NO_x^{out})_L \tag{16}$$

where $(NO^{in})_L$ is the total amount of NO fed during the lean period and $(NO_x^{out})_L$ is the total amount of NO and NO_2 leaving the reactor during the same period. These amounts correspond to the areas graphically represented in Fig. 8, which can be calculated by the corresponding numerical integrations.

When the cumulative NO_x trapped (eqn. 16) is expressed as a percentage of the NO_x fed, then it is referred to as the NO_x storage capacity,

$$\mu_{STO} = \frac{NO_x^{stored}}{(NO^{in})_L} \times 100 \tag{17}$$

The catalyst performance during the rich period was described, on the one hand, by the reduction conversion (X_R) defined as the percentage of NO_x reduced over the total amount of NO_x to be reduced. The latter accounts for the sum of the NO_x stored during the lean period plus the NO continuously fed during the rich period. Then,

$$X_R(\%)=\frac{NO_x^{reduced}}{NO_x^{to\ be\ reduced}} \times 100 = \frac{\left[NO_x^{stored}+(NO^{in})_R \right]-(NO_x^{out})_R}{\left[NO_x^{stored}+(NO^{in})_R \right]} \times 100 \tag{18}$$

On the other hand, the NO_x reduction conversion needs to be complemented with the selectivity to different nitrogen species, including dinitrogen oxide, ammonia and nitrogen. Then, selectivities were defined as

$$S_{NH_3} = \frac{NH_3^{out}}{NH_3^{out}+2N_2^{out}+2N_2O^{out}} \times 100 \tag{19}$$

$$S_{N_2O} = \frac{2N_2O^{out}}{NH_3^{out}+2N_2^{out}+2N_2O^{out}} \times 100 \tag{20}$$

Fig. 8 shows over a model profile for NO_x and NH_3 molar flows at the exit of the reactor the areas corresponding to the component amounts leaving the reactor during the regeneration period. The N_2O amount leaving the reactor is calculated similarly from the corresponding outlet profile. The FTIR technique is not able to analyze N_2 so that the amount of this component should be calculated from the nitrogen mole balance

$$NO_x^{stored}+(NO^{in})_R=NH_3^{out}+2N_2^{out}+2N_2O^{out}+(NO_x^{out})_R \tag{21}$$

i.e. the amount of NO_x stored plus the NO_x amount fed during the lean period equals the NO_x amount leaving the reactor and those amounts converted into NH_3, N_2 and N_2O. Then, the selectivity to N_2 can be expressed as

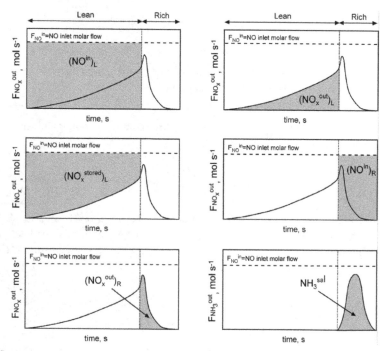

Figure 8. Graphical representation of areas corresponding to magnitudes needed to define parameters for evaluation of the catalyst performance in the NSR process.

$$S_{N_2} = \frac{2N_2}{NH_3^{out} + 2N_2^{out} + 2N_2O^{out}} \times 100 =$$

$$\frac{\left[NO_x^{stored} + (NO^{in})_R \right] - \left[NH_3^{out} + 2N_2O^{out} + (NO_x^{out})_R \right]}{NH_3^{out} + 2N_2^{out} + 2N_2O^{out}} \times 100 \qquad (22)$$

On the other hand, the nitrogen mole balance can be checked if one can determine the amount of N_2 at the reactor exit by the adequate analysis technique, e.g. quantitative mass spectrometry.

The parameters above defined are useful to compare independently the NO_x storage capacity during the lean period and the NO_x reduction conversion during the rich period. The N_2/NH_3 selectivities have been calculated averaged over the whole cycle, as peaks corresponding to those compounds can be seen at the outlet during the rich time but also continuing during the subsequent lean period (see Fig. 9b later). In the case of a conventional Pt–BaO/Al₂O₃ NSR system, the catalyst should operate to exhibit high NO_x storage capacity with also high selectivity to N_2. Thus, definition of a single parameter giving information of the trap performance over the whole storage-reduction cycle would be very convenient to know how efficiently the NSR system is running. This global parameter should take into account the storage capacity, the reduction conversion and the selectivity of the reaction, giving a general vision of the efficiency of the whole NSR process. Thus, the global NSR efficiency, referred to the N_2 production over the total amount of NO_x fed, can be calculated as

$$\varepsilon_{NSR} = \frac{2N_2^{out}}{(NO^{in})_L + (NO^{in})_R} \times 100 = \frac{2N_2^{out}}{NH_3^{out} + 2N_2^{out} + 2N_2O^{out} + (NO_x^{out})_R} \times 100 \qquad (23)$$

5.2. Optimal control of the NSR technology by managing the amount of reductant injected during the regeneration period

NO_x storage-reduction experiments were carried out in a downflow steel reactor. The monolithic catalyst (25 mm in length and diameter, 3.5 g) was placed in the bottom part of the reactor and the set was introduced in a 3-zone oven. The temperature at the entry and exit of the reactor was continuously monitored. The experimental conditions are shown in Table 4. The feedstream during storage was 380 ppm NO/6% O_2/N_2. Gases were fed through mass controllers with a total volumetric flow of 3,365 l min^{-1}, corresponding to a GHSV of 32,000 h^{-1} (STP).

The problem was stated as follows: to find the values of the operational variables, including lasting time of the storage period (lean mixture), lasting time of the regeneration period (rich mixture) and hydrogen concentration injected during the regeneration period, which allow to achieve the maximum NSR efficiency (Eq. 23).

Operational parameters	Values
Temperature, °C	330
Total volumetric flow, l min^{-1} (STP)	3.365
Spatial velocity, GHSV, h^{-1}	32,100
Lean period time duration, s	145, 290, 595
Rich period time duration, s	16 - 47
Lean mixture composition	380 ppm NO, 6% O_2, N_2 to balance
Rich mixture composition	380 ppm NO, 0.41 – 2.36% H_2, N_2 to balance

Table 4. Experimental conditions.

5.2.1. Effect of the H_2 concentration in the regeneration stream on NO_x storage and reduction

Fig. 9 shows the effect of hydrogen concentration in the rich stream on the NO_x and NH_3 concentrations at the exit of the reactor, for experiments carried out at 330 °C. During the lean period the stream composition was 6% O_2 and 380 ppm NO, with N_2 to balance. After 145 s of lean period the oxygen was shifted to hydrogen at different concentrations (0.79, 1.1 and 2.32%), maintaining 380 ppm NO in the feedstream for the regeneration period of 25 s.

The evolution of NO_x concentration at the reactor exit during the storage and regeneration periods is that typical for the NSR process (see Fig. 2), i.e. at the beginning of the lean period all amount of NO_x is stored, and this amount is gradually reduced as the adsorption sites are being saturated, then increasing the NO_x exiting the reactor. It can be seen in Fig. 9a, nevertheless, that the NO_x concentration at the end of the lean period (145 s) did not reach the initial concentration (380 ppm NO), i.e. the catalyst was not completely saturated. The monitored values resulted in 240, 170 and 145 ppm NO_x for runs 1, 2 and 3, respectively.

After 145 s of lean period, the shifting of oxygen by hydrogen to the entry of the reactor provokes the release of the previously stored NO_x which is eventually reduced to N_2O, NH_3 and N_2, according to the mechanisms explained in section 4. The evolution of ammonia concentration at the reactor exit is shown in Fig. 9b. At 330°C, however, N_2O was not practically apreciated at the reactor exit.

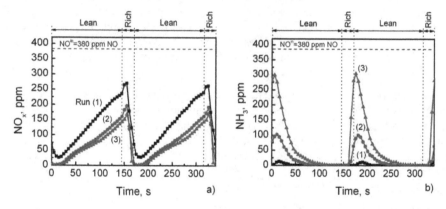

Figure 9. Concentration of (a) NO_x and (b) NH_3 at the reactor exit for two consecutive storage-reduction, for different H_2 concentration during the rich period: (1) 0.79% H_2; (2) 1.1% H_2; (3) 2.32% H_2. t_L=145 s, t_R=25 s.

The hydrogen concentration influences significantly the formation of ammonia at the exit, from 15 ppm NH_3 for 0.79% H_2 to a maximum of 300 ppm NH_3 for 2.32% H_2. In fact, the higher H_2 concentration in the reduction stream the higher NH_3 concentration at the exit, as previously reported [56,78,87-89,91]. These experiments also confirmed the delay in the ammonia detection at the reactor exit related to the beginning of the rich period, as it was observed in the experiments of section 3.

The amount of hydrogen supplied during the rich period also influences the storage capacity and distribution of products. In fact, the storage capacity (μ_{STO}, eqn.17) increased with H_2 concentration, resulting in 40.8, 77.3 y 80.5% for runs 1, 2 and 3, respectively. Similarly, selectivity to NH_3 (S_{NH_3}, eqn. 19) also increased with H_2 concentration, resulting in 2, 10 and 29%, respectively. This influence can be better observed in Fig. 10, where values of the response variables corresponding to additional H_2 concentration experiments have been included. The response variables calculated have been: storage capacity (μ_{STO}, eqn. 17), NOₓ reduction conversion (X_R, eqn. 18), selectivity to N₂ (S_{N_2}, eqn. 22) and NSR efficiency (ε_{NSR}, eqn.23).

Fig. 10a evidences a linear increase of the storage capacity (red line) with H₂ concentration up to 1.1%. Above 1.1% H₂, the storage capacity was maintained almost constant in about 80%. This evolution can be explained from the regeneration mechanism proposed in section 4 [84,85]. During the regeneration step there exits a hydrogen front that travels along the catalyst while regenerating the adsorption sites. If the amount of hydrogen fed is enough, the regeneration front will travel until regeneration of the complete trap. On the contrary, if the reductant is in defect, the regeneration front will not arrive to the final part of the reactor, and the adsorption sites downstream could not be regenerated. This explains that 0.79% H₂ was not able to regenerate the whole catalyst resulting in a limited storage capacity (40.8%). While increasing the H₂ concentration, the regeneration front was able to reach more advanced positions and thus regenerates more adsorption sites, resulting in higher storage capacity during the lean period. Above 1.1% H₂ it is assumed that the regeneration front travels the whole catalyst with total regeneration of the trap. Thus, higher H₂ concentration did not produce significant variations of the storage capacity. Another explanation was supplied by Clayton et al. [86] based on the different storage regions related to the barium phase storage sites based on their proximity to Pt crystallites. They associated the extension of these regions with the platinum dispersion and the reaction temperature.

The reduction conversion (blue line) follows a similar trend to the storage capacity (Fig. 10a). Hydrogen concentration above 1.1% resulted in almost total conversion (97%), whereas lower concentrations resulted also in lower conversion level. In fact, when the supply of H₂ is not enough to allow the hydrogen front achieving the complete regeneration of the trap, e.g. 0.79% H₂, the NOₓ reduced/NOₓ released ratio is increasing, then decreasing the NOₓ reduction conversion to 85%.

Concerning selectivity N₂/NH₃ (remember that N₂O concentration was negligible at 330 °C), Fig. 10a shows that formation of ammonia decreased with lower hydrogen concentration during the rich period, being very low with 0.79% H₂. Thus, when the catalyst regeneration was carried out under low hydrogen concentration (0.79% H₂) the selectivity to N₂ is practically total and only 15 ppm NH₃ were detected. When increasing the H₂ concentration, the selectivity to N₂ decreased progressively in favour of ammonia (Fig. 9b). This agrees with the observation of Clayton et al. [91] that nitrogen selectivity increased with the NOₓ/H₂ ratio.

The trends of NOₓ storage capacity, NOₓ conversion and N₂/NH₃ selectivity above explained, suggest differentiation of two different zones in Fig. 10, limited to each other by 1.1% H₂. In zone A occurs that hydrogen is the reactant that limits the reduction of NOₓ, and complete NOₓ reduction cannot be achieved (Fig. 8, 0.79% H₂). On the contrary, in zone B the reaction occurs with excess of hydrogen and NOₓ are completely reduced. This is also in agreement with the fact that in zone A the NOₓ storage capacity is limited as all barium sites cannot be regenerated, whereas in zone B the excess of hydrogen enhances the ammonia formation.

Fig. 10b shows the evaluation of NSR efficiency (ε_{NSR}) with H_2 concentration. As already mentioned, this can be considered as a global parameter that considers the complete storage-reduction cycle as determining the molar amount of nitrogen at the reactor exit over the molar amount of NO at the entry, expressed as percentage. In fact, with some simple mathematical rearrangements of eqns. (17)-(23), the relationship between the NSR efficiency and the previous response variables can be found, which is expressed as:

$$\varepsilon_{NSR} = X_R S_{N_2} (\mu_{STO} \tau_L + \tau_R) \qquad (24)$$

where τ_L and τ_R are the dimensionless lean and rich times

$$\tau_L = \frac{t_L}{t_L + t_R} \qquad\qquad \tau_R = \frac{t_R}{t_L + t_R} \qquad (25)$$

The opposite trend shown by S_{N_2} and μ_{STO} with the amount of hydrogen fed during the rich period (Fig. 10a), makes the NSR efficency to reach a maximum at some intermediate value of $\%H_2$ (eqn. 17), as seen in Fig. 10b. With low $\%H_2$ (zone A) high S_{N_2} is achieved but μ_{STO} is limited, whereas with high $\%H_2$ (zone B) low S_{N_2} (high formation of NH_3) is achieved but μ_{STO} is maintained maximum. Then, the maximum of ε_{NSR} is achieved for 1.1% H_2, just the border between zones A and B, where the amount of hydrogen is that needed to make the complete regeneration of barium sites but not more to avoid formation of ammonia.

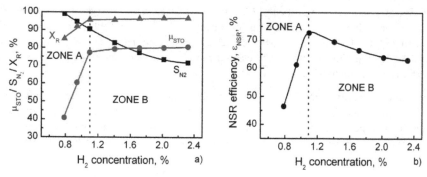

Figure 10. (a) Evolution of the NO_x storage capacity, reduction conversion and selectivity to nitrogen as a function of H_2 concentration during regeneration. (b) NSR efficiency vs. H_2 concentration.

5.2.2. Influence of storage and regeneration period duration on the NO_x storage and reduction

In the previous section all experiments were performed with same duration of the lean and rich periods, i.e. t_L=145 s and t_R=25 s, and varying only the H_2 concentration during the rich period. On the other hand, it can be concluded that the total amount of hydrogen fed during the rich period determines the maximum NSR efficiency. Obviously this amount of hydrogen can be considered proportional to the product $C_{H_2} \times t_R$, so that it can also be varied by modifying the duration of the rich period. Pereda–Ayo et al. [56] made experiments looking for different combinations of pairs (C_{H_2}, t_R) that achieved maximum NSR efficiency, when the lean period duration was maintained in 145 s. The results of these experiments can be represented as the locus of all these combinations as shown in Fig. 11 (t_L=145 s, red curve) and defines the isocurve of operational conditions to carry out the global NSR process efficiently.

The shape of the isocurve represented in Fig. 11 indicates the inverse relationship between the regeneration time and the reductant concentration to achieve an efficient NSR process, i.e. the shorter reduction time, the higher reductant concentration needed to achieve maximum efficiency. This finding implies again that the supply of the reductant H₂ is controlling the NSR process. This was also observed by Mulla et al. [81] that measured the time required for regenerate the trap catalyst by the width of the N₂ pulse in a mass spectrometer. Analogously, Nova et al. [45] had also noticed that the N₂ production during the regeneration of a Pt–BaO/Al₂O₃ catalyst was limited by the amount of H₂ fed to the reactor.

5.2.3. Extension of the duration of lean period on the NSR performance

To further investigate the optimal conditions to operate the NSR process efficiently, the duration of the storage period was varied, extending the lean period duration from 145 to 290 and 595 s [56]. Again, analogous NO$_x$ storage and reduction experiments were performed with those extended times and the same protocol as before. It has been verified that the same storage capacity was obtained for a given lean period duration, resulting in additional isocurves shown in Fig. 11. As expected, the NO$_x$ storage capacity decreased as the lean period duration increased. When the lean period is longer the catalyst is closer to the saturation level, and consequently the NO$_x$ storage capacity decreases, i.e. 77, 55 and 35% for t_L = 145, 290 and 595 s, respectively. As for selectivity to nitrogen, very similar values were found around 90%, irrespective of the studied variables (C_{H_2}, t_R, t_L), provided that the operation is achieved with maximum efficiency (all points in every isocurve in Fig. 11).

As noted above, the nitrogen production during the regeneration of the catalyst is limited by the amount of hydrogen fed to the reactor. Likewise, increasing the hydrogen supply rate is expected to have a linear effect on the overall rate of NO$_x$ reduction. For our experiments, Fig. 12 shows a linear effect of the hydrogen concentration fed during the rich period on the overall NO$_x$ reduction rate, independent of the duration of the lean period, thus suggesting again that the regeneration step is limited by the amount of hydrogen fed. The linear relationship implies that the time required for complete regeneration should be inversely proportional to the reductant amount of hydrogen fed, as shown in the isocurves of Fig. 11. Mulla et al. [81] also reported the overall rate for NO$_x$ reduction as a linear function of rate of flow of H-atoms in the form of H₂ or NH₃ at 300 °C. Their observations also confirmed that the regeneration process was not mass transfer or kinetically limited, but it was controlled by the supply of the reductant H₂.

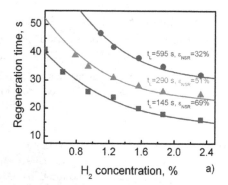

Figure 11. Operation map of the Pt-Ba/Al₂O monolithic catalyst. Relationship between operational variables (C_{H_2}, t_P, t_R) for carrying out the NSR process efficiently.

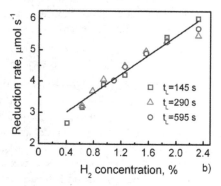

Figure 12. Linear relation between overall NOₓ reduction rate *vs.* hydrogen concentraton fed during rich period.

Finally, the term "operation map" is suggested for the set of curves represented in Fig. 11, as a tool for finding any combination of the three studied operational variables: the duration of the lean period, the duration of the rich period and the concentration of the reducing agent: to run the NSR process efficiently. Two ideas may arise from this map. First, every manufactured catalyst can be associated with its own map, so that the comparison of maps will provide information about their relative efficiency when running under the real application. Secondly, one may wonder if any operation point in the map of Fig. 11 is susceptible to be chosen as the best combination (C_{H_2}, t_R, t_L) to run in real application.

5.3. Performance of NOₓ storage–reduction catalyst in the temperature–reductant concentration domain by response surface methodology

All previous experiments were carried out at the temperature of 330 °C, at which N₂O at the reactor·exit was negligible, thus being selectivity distributed between N₂ and NH₃, depending on the lean and rich period durations and the hydrogen concentration during the rich period. In this section, the NSR performance trends of the Pt–Ba/Al₂O₃ monolith catalyst will be studied at different temperatures and varying the hydrogen concentration fed during the regeneration period by the response surface methodology (RSM). The NO_x storage and reduction behaviour was tested over 9 levels of temperature: 100, 140, 180, 220, 260, 300, 340, 380 and 420 °C and 9 levels of hydrogen concentration: 0.4, 0.55, 0.7, 0.85, 1, 1.5, 2, 2.5 and 3% [57].

Fig. 13a shows the NO_x storage capacity (μ_{STO}) response surface in the hydrogen concentration and reactor inlet temperature domain. With the aim of finding the optimal region, isocurves corresponding to different levels of μ_{STO} projected to the $T-C_{H_2}$ space are drawn in Fig. 13b. In the region comprised between temperatures of 220 and 260 °C and hydrogen concentrations of 1.75 and 3% a nearly flat surface corresponding to the maximum NO_x storage capacity above 80% is observed (shaded region). These optimal operational conditions correspond to intermediate temperature and excess of hydrogen. At lower temperatures the conversion of NO to NO₂ was not favoured whereas at higher temperatures the stability of the stored nitrates was reduced leading in both cases to a decrease in the NO_x storage capacity [85]. On the other hand, operating with low hydrogen concentration (<1% H₂) resulted also in a sharp decrease in the NO_x storage capacity due to the incomplete regeneration of the catalyst [56].

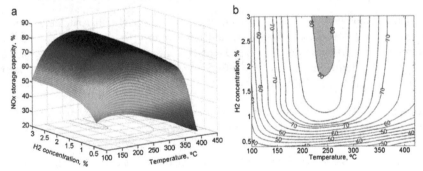

Figure 13. (a) NO_x storage capacity response surface in the temperature and hydrogen dose domain, and (b) isocurves corresponding to different levels of trapping efficiencies projected to the T-CH₂ space.

Likewise, Fig. 14a shows the selectivity to nitrogen response surface in the hydrogen concentration and temperature domain and Fig. 14b the projected iso-selectivity to nitrogen curves. At low temperature (<150 °C) the selectivity to nitrogen resulted nearly independent of the hydrogen concentration as almost vertical lines can be observed. In this region, the product selectivity changed from N₂O at low H₂ concentrations to NH₃ at higher ones, but remaining practically constant the selectivity to nitrogen. For example, at 100 °C, N₂O/NH₃/N₂ = 54.5/5.1/40.4 for 0.4% H₂; N₂O/NH₃/N₂ = 34.7/33.1/33.2 for 1% H₂; N₂O/NH₃/N₂ = 19.5/48.8/31.7 for 3% H₂. For higher temperatures (>180 °C), where the formation of N₂O was negligible, the influence of hydrogen concentration on the selectivity to nitrogen became markedly significant. In this region, the higher hydrogen concentration the lower nitrogen selectivity, and therefore the higher ammonia, was obtained. For example, at 340 °C, $N_2O/NH_3/N_2 = 3.4/2.2/94.5$ for 0.4% H_2; $N_2O/NH_3/N_2 = 0.8/9.4/89.9$ for 1% H_2; $N_2O/NH_3/N_2 = 0.6/25.0/74.4$ for 3% H_2.

The optimal operational window which resulted in a selectivity to nitrogen higher than 90% was situated at intermediate-high temperatures ($T > 250$ °C) and low hydrogen concentrations ($C_{H_2} < 1\%$) as it can be seen in Fig. 14b (shaded region).

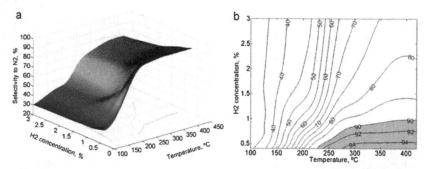

Figure 14. (a) Selectivity to N_2 response surface in the temperature and hydrogen dose domain, and (b) isocurves corresponding to different levels of nitrogen selectivity projected to the T–C_{H2} space.

The optimal operating region for maximizing NO_x trapping efficiency and nitrogen selectivity, Fig. 13b and 14b, respectively, did not intercept to each other. The first was maximum at intermediate temperatures and high H_2 concentrations, $T = 220$–260 °C and C_{H2} = 1.75–3% H_2; whereas the selectivity to nitrogen was favoured by high temperatures and low H_2 concentrations, $T > 250$ °C and $C_{H_2} < 1\%$.

In conventional NSR systems NO_x conversion towards N_2 should be maximized while NH_3 formation should be avoided. The percentage of NO_x converted into nitrogen relative to the total amount of NO entering the trap has been defined as global NSR efficiency (ε_{NSR}, eqn. 23), which allows one to look for the optimal combination of reactor inlet temperature and hydrogen concentration during rich period to obtain the most efficient NSR operation. Fig. 15a shows the ε_{NSR} surface response in the hydrogen concentration and temperature domain and Fig. 15b the projected iso-efficiency curves. As it could be expected, an optimal interval of temperature and hydrogen concentration situated between the optimal operational conditions to obtain maximum storage and maximum selectivity was found. The NSR efficiency resulted higher than 60%, that is, more than 60% of NO entering the trap was converted into nitrogen, in the following operating window: $T = 250$–350 °C and C_{H_2} = 0.8–1.5. This region provides the best compromise between NO_x storage capacity and selectivity to N_2 to maximize the nitrogen production at the reactor exit related to the total amount of NO at the reactor inlet. The most efficient operation corresponds to 0.9% H_2 and 280 °C reaching 65% of global NSR efficiency (Fig. 15b).

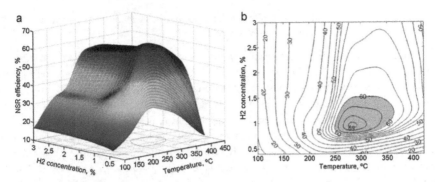

Figure 15. (a) Global NSR efficiency response surface in the temperature and hydrogen dose domain, and (b) isocurves corresponding to different levels of NSR efficiencies projected to the T–C_{H2} space.

6. Conclusions

The NO_x storage and reduction technology for diesel eshaust aftertreatment has been studied in the present chapter, including the synthesis of Pt-BaO/Al_2O_3 monolith catalyst, the involved reaction mechanisms or chemistry of the process, and the control of engineering parameters, of importance in the real application in automobiles, to remove nitrogen oxides most efficiently by conversion to nitrogen.

The preparation methodology for Pt-BaO/Al_2O_3 monolith catalyst has been described. The monolith is primarily washcoated with a thin film of porous alumina. Then platinum is incorporated by adsorption (ion exchange) from a $Pt(NH_3)_4(NO_3)_2$ aqueous solution. Finally, the barium as NO_x storage component is incorporated by dry impregnation from a $Ba(CH_3-COO)_2$ aqueous solution. This procedure achieves homogeneous distribution as well as high dispersion of platinum and barium on the catalyst surface, then providing the adequate Pt-Ba proximity which enhances the interaction needed for Pt to promote both the initial NO oxidation and the reduction of N_xO_y adspecies on the Ba sites.

The chemistry of NO_x regeneration and reduction mechanisms has been reviewed. Operando FTIR experiments of NO_x adsorption on powder Pt-Ba/Al_2O_3 samples has shown that below 250 °C nitrite species are predominant whereas above 250 °C nitrate species are predominant. Thus, two parallel routes have been verified. The "nitrite route" where NO is oxidized on Pt sites and stored onto Ba neighbouring sites in the form of nitrite ad-species which can progressively transform into nitrates depending on the reaction temperature. The second route, called "nitrate route" implies the oxidation of NO to NO_2 on Pt, then NO_2 desproportionation on Ba to form nitrates NO evolved into the gas phase.

During the regeneration period, when oxygen is shifted by hydrogen, the reduction of stored nitrites and nitrates leads to the formation of different nitrogen containing species, namely N_2, N_2O and NH_3 along with water. Nitrogen formation involves first the fast formation of ammonia by reaction of nitrates with H_2 and then the subsequent conversion of the ammonia formed with stored nitrates leading to the selective formation of N_2. At temperature above 330 °C, N_2O was almost negligible.

In the automobile practice, the operational conditions at which the process is conducted affect significantly the NSR behaviour of a Pt-BaO/Al$_2$O$_3$ monolith catalyst, such as the duration of lean and rich periods and the concentration of reductant fed during the regeneration period. There exists a given amount of hydrogen which is needed to achieve the complete reduction of NO$_x$ (stored during the lean period and fed during the rich period). Below that minimum, as regeneration is not complete the reduction conversion is lower and consequently the storage capacity in the subsequent lean period is also reduced. However, with hydrogen in defect very high selectivity towards N_2 is achieved. On the other hand, with hydrogen in excess the formation of ammonia increases notably, although NO$_x$ storage capacity is practically maintained at maximum and almost total reduction conversion is achieved. The maximum global NSR efficiency (percentage of N_2 at the exit related to NO at the entry) is achieved just at the stoichiometric point, when the amount of H_2 is neither in defect nor in excess. The amount of hydrogen fed during the rich period is proportional to the product $C_{H_2} \times t_R$ so that this amount can be controlled by managing either H_2 concentration or duration of the regeneration period. In fact, there exist different combinations (C_{H_2}, t_R) which achieve similar NSR efficiency. The locus of these combinations conforms the isoeffiency curve map with NSR efficiency as the response parameter.

The combined analysis of temperature (100-420 °C) and H_2 concentration (0.4-3%), maintaining lean and rich period times in 145 and 25 s respectively, has allowed to find the maximum storage capacity at intermediate temperature (~240 °C) and high reductant concentration (>2% H_2). Maximum selectivity to N_2 has been obtained operating at high temperature (>300 °C) and hydrogen in defect (<1% H_2). The optimal control is performed at intermediate position, i.e. 270 °C and 1% H_2, at which the maximum global NSR efficieny is achieved.

Nomenclature

Abreviations

ADS   Adsorption from solution, catalyst preparation procedure.
DI   Dry impregnation
FTIR   Fourier Transform Infrared spectroscopy.
ICP-MS   Inductively Coupled Plasma Mass Spectroscopy.
LNT   Lean NO$_x$ Trap.
MS   Mass Spectroscopy.
NSR   NO$_x$ Storage and Reduction.
PM   Particulate matter.
RSM   Response Surface Methodology.
SEM-EDX   Scanning Electronic Microscopy-Energy Dispersed X-Ray Spectroscopy.
SCR   Selective Catalytic Reduction.

STP	Standard temperature and pressure
TEM	Transmission Electronic Microscopy.
TWC	Three way catalyst.
WI	Wet impregnation, catalyst preparation procedure.

Variables

A/F	Air-to-fuel ratio.
C_{H_2}	Concentration of the reductant agent (hydrogen) in the regeneration feedstream, %.
NO_x	Nitrogen oxides ($NO+NO_2$).
NO_x^{stored}	Amount of NO_x stored, mol s^{-1}.
$(NO^{in})_L$	Total amount of NO fed to the system during the duration of lean period (storage), mol.
$(NO_x^{out})_L$	Total amount of NO_x at the exit of reactor during the duration of lean period (storage), mol.
$(NO^{in})_R$	Total amount of NO fed to the system during the duration of rich period (regeneration), mol.
$(NO_x^{out})_R$	Total amount of NO_x at the exit of reactor during the duration of rich period (regeneration), mol.
S_{N_2}	Selectivity towards nitrogen, eqn. (22), %.
S_{N_2O}	Selectivity towards N₂O, eqn. (20), %.
S_{NH_3}	Selectividad towards ammonia, eqn. (19), %.
T	Temperature, °C.
t_L	Lasting time of lean period (storage), s.
t_R	Lasting time of rich period (regeneration), s.
X_R	NO_x conversion during the regeneration period, eqn. (18), %.

Greek symbols

ε_{NSR}	Global NSR efficiency, eqn. (23) , %.
μ_{STO}	NO_x storage capacity, eqn. (17), %.
τ_L	dimensionless lean period time
τ_R	dimensionless rich period time

Author details

Beñat Pereda-Ayo and Juan R. González-Velasco*
Department of Chemical Engineering, Faculty of Science and Technology, University of the Basque Country UPV/EHU, Bilbao, Spain

---

* Corresponding Author

Acknowledgement

The authors wish to acknowledge the financial support provided by the Spanish Science and Innovation Ministry (CTQ2009-125117) and the Basque Government (Consolidated Research Group, GIC 07/67-JT-450-07).

7. References

[1] Pereda-Ayo B. NO_x storage and reduction (NSR) for diesel engines: synthesis of Pt-Ba/Al_2O_3 monolith catalyst, reaction mechanisms and optimal control of the process. PhD Thesis, University of the Basque Country, Bilbao, Spain; 2012.

[2] Mazzarella G, Ferraraccio F, Prati MV, Annunziata S, Bianco A, Mezzogiorno A, Liguori G, Angelillo IF, Cazzola M. Effects of diesel exhaust particles on human lung epithelial cells: An in vitro study. Respiratory Medicine 2007;101 1155-1162.

[3] Derwent RG. The long-range transport of ozone within Europe and its control. Environmental Pollution 1990; 299-318.

[4] Parvulescu VI, Grange P, Delmon B. Catalytic removal of NO. Catalysis Today 1998;46 233-316.

[5] Folinsbee LJ. Human health-effects of air-pollution. Environmental Health Perspectives 1993;100 45-56.

[6] Krupa SV, Kickert RN. The greenhouse effect - impacts of ultraviolet-B (Uv-B) radiation, carbon-dioxide (CO_2), and ozone (O_3) on vegetation. Environmental Pollution 1989;61 263-393.

[7] Klingstedt F, Arve K, Eranen K, Murzin DY. Toward improved catalytic low-temperature NO_x removal in diesel-powered vehicles. Accounts of Chemical Research 2006;39 273-282.

[8] Dieselnet. Emission standards. European Union. Cars and light trucks. http://www.dieselnet.com/standards/eu/ld.php (accessed 1 May 2012).

[9] Taylor KC. Nitric-oxide catalysis in automotive exhaust systems. Catalysis Reviews Science Eng. 1993;35 457-481.

[10] Tamaru K, Mills GA. Catalysts for control of exhaust emissions. Catalysis Today 1994;22 349-360.

[11] Heck RM, Farrauto, RJ. Automobile exhaust catalysts. Applied Catalysis A: General 2001;221 443-457.

[12] González-Velasco JR, Entrena J, González-Marcos JA, Gutiérrez-Ortiz JI, Gutiérrez-Ortiz MA. Preparation, activity and durability of promoted platinum catalysts for automotive exhaust control. Applied Catalysis B: Environmental 1994;3 191-204.

[13] Numan JG, Robota HJ, Cohn MJ, Bradley SA. Physicochemical properties of Ce-containing three-way catalysts and the effect of Ce on catalyst activity. Journal of Catalysis 1992;133 309-324.

[14] Engler BH, Lindner D, Lox ES, Schäfer-Sindlinger, Ostgathe K. Development of improved Pd-only and Pd/Rh three-way catalysts. Studies in Surface Science and Catalysis 1995;96 441-460.

[15] Heck RM, Farrauto RJ, Gulati S. Catalytic Air Pollution Control: Commercial Technology. New Yersey: John Wiley & Sons; 2009.

[16] Seijger GBF. Cerium-ferrierite catalyst systems for reduction of NO$_x$ in lean burn engine exhaust gas. PhD Thesis, Technical University Delft, Delft, Netherlands; 2002.

[17] González-Velasco JR, González-Marcos MP, Gutiérrez-Ortiz MA, Botas-Echevarría JA. Catálisis, automóvil y medio ambiente. Anales de Química 2002;4 24-35.

[18] Basile F, Fomasari G, Grimandi A, Livi M, Vaccari A. Effect of Mg, Ca and Ba on the Pt-catalyst for NOx storage reduction. Applied Catalysis B: Environmental 2006;69 58-64.

[19] Trichard JM. Current tasks and challenges for exhaust after-treatment research: An industrial viewpoint. Studies in Surface Science and Catalysis 2007;171 211-233.

[20] Centi G, Perathoner S. Introduction: State of the art in the development of catalytic processes for the selective catalytic reduction of NO$_x$ into N$_2$. Studies in Surface Science and Catalysis 2007;171 1-23.

[21] Johnson TV. Review of diesel emissions and control. International Journal of Engine Research 2009;10 275-285.

[22] Maricq MM. Chemical characterization of particulate emissions from diesel engines: A review. Journal of Aerosol Science 2007;38 1079-1118.

[23] Fino D. Diesel emission control: Catalytic filters for particulate removal. Science and Technology of Advanced Materials 2007;8 93-100.

[24] Biswas S, Verma V, Schauer JJ, Sioutas C. Chemical speciation of PM emissions from heavy-duty diesel vehicles equipped with diesel particulate filter (DPF) and selective catalytic reduction (SCR) retrofits. Atmospheric Environment 2009;43 1917-1925.

[25] Brandenberger S, Krocher O, Tissler A, Althoff R. The state of the art in selective catalytic reduction of NO$_x$ by ammonia using metal-exchanged zeolite catalysts. Catalysis Reviews, Science and Engineering 2008;50 492-531.

[26] Amiridis MD, Zhang TJ, Farrauto RJ. Selective catalytic reduction of nitric oxide by hydrocarbons. Applied Catalysis B: Environmental 1996;10 203-227.

[27] Takahashi N, Shinjoh H, Iijima T, Suzuki T, Yamazaki K, Yokota K, Suzuki H, Miyoshi N, Matsumoto S, Tanizawa T, Tanaka T, Tateishi S, Kasahara K. The new concept 3-way catalyst for automotive lean-burn engine: NO$_x$ storage and reduction catalyst. Catalysis Today 1996;27 63-69.

[28] Matsumoto SI. Recent advances in automobile exhaust catalysts. Catalysis Today 2004;90 183-190.

[29] Belton DN, Taylor KC. Automobile exhaust emission control by catalysts. Current Opinion in Solid State & Materials Science 1999;4 97-102.

[30] Alkemade UG, Schumann B. Engines and exhaust after treatment systems for future automotive applications. Solid State Ionics 2006;177 2291-2296.

[31] Twigg MV. Progress and future challenges in controlling automotive exhaust gas emissions. Applied Catalysis B: Environmental 2007;70 2-15.

[32] Fridell E., Skoglundh M., Johansson S., Westerberg BR, Törncrona A, Smedler G. Investigations of NOₓ storage catalysts. Studies in Surface Science and Catalysis 1998;116 537-547.

[33] Miyoshi N, Matsumoto S, Katoh T, Tanaka T, Harada J, Takahashi N, Yokota K, Sugiara M, Kasahara K. Development of new concept three-way catalyst for automotive lean-burn engines. SAE Technical Paper Series 950809; 1995.

[34] Epling WS, Campbell LE, Yezerets A, Currier NW, Parks JE. Overview of the fundamental reactions and degradation mechanisms of NOₓ storage/reduction catalysts. Catalysis Reviews, Science and Engineering 2004;46 163-245.

[35] Roy S, Baiker A. NOₓ Storage-reduction catalysis: from mechanism and materials properties to storage-reduction performance. Chemical Reviews 2009;109 4054-4091.

[36] Liu G, Gao PX. A review on NOₓ storage/reduction catalysts: mechanism, materials and degradation studies. Catalysis Science & Technology 2011;1 552-568.

[37] Li YJ, Roth S, Dettling J, Beutel T. Effects of lean/rich timing and nature of reductant on the performance of a NOₓ trap catalyst. Topics in Catalysis 2001;16 139-144.

[38] Epling WS, Yezerets A, Currier NW. The effect of exothermic reactions during regeneration on the NOₓ trapping efficiency of a NOₓ storage/reduction catalyst. Catalysis Letters 2006;110 143-148.

[39] Abdulhamid H, Fridell E, Skoglundh M. The reduction phase in NOₓ storage catalysis: Effect of type of precious metal and reducing agent. Applied Catalysis B: Environmental 2006;62 319-328.

[40] Meille V. Review on methods to deposit catalysts on structured surfaces. Applied Catalysis A: General 2006;315 1-17.

[41] Nijhuis TA, Beers AEW, Vergunst T, Hoek I, Kapteijn F, Moulijn JA. Preparation of monolithic catalysts. Catalysis Reviews, Science and Engineering 2001;43 345-380.

[42] Avila P, Montes M, Miro E. Monolithic reactors for environmental applications - A review on preparation technologies. Chemical Engineering Journal 2005;109 11-36.

[43] Piacentini M, Maciejewski M, Baiker A. NOₓ storage-reduction behavior of Pt-Ba/MO₂ (MO₂ = SiO₂, CeO₂, ZrO₂) catalysts. Applied Catalysis B: Environmental 2007;72 105-117.

[44] Malpartida I, Vargas MAL, Alemany LJ, Finocchio E, Busca G. Pt-Ba-Al₂O₃ for NOₓ storage and reduction: Characterization of the dispersed species. Applied Catalysis B: Environmental 2008;80 214-225.

[45] Nova I, Lietti L, Forzatti P. Mechanistic aspects of the reduction of stored NOₓ over Pt-Ba/Al₂O₃ lean NOₓ trap systems. Catalysis Today 2008;136 128-135.

[46] Agrafiotis C, Tsetsekou A. The effect of powder characteristics on washcoat duality. Part I: Alumina washcoats. Journal of the European Ceramic Society 2000;20 815-824.

[47] Pereda-Ayo B, López-Fonseca R, González-Velasco JR. Influence of the preparation procedure of NSR monolithic catalysts on the Pt-Ba dispersion and distribution. Applied Catalysis A: General 2009;363 73-80.

[48] Tsetsekou A, Agrafiotis C, Milias A. Optimization of the rheological properties of alumina slurries for ceramic processing applications - Part I: Slip-casting. Journal of the European Ceramic Society 2001;21 363-373.

[49] Valentini M, Groppi G, Cristiani C, Levi M, Tronconi E, Forzatti P. The deposition of gamma-Al2O3 layers on ceramic and metallic supports for the preparation of structured catalysts. Catalysis Today 2001;69 307-314.

[50] Agrafiotis C, Tsetsekou A. Deposition of meso-porous gamma-alumina coatings on ceramic honeycombs by sol-gel methods. Journal of the European Ceramic Society 2002;22 423-434.

[51] Lindholm A, Currier NW, Dawody J, Hidayat A, Li J, Yezerets A, Olsson L. The influence of the preparation procedure on the storage and regeneration behavior of Pt and Ba based NOx storage and reduction catalysts. Applied Catalysis B: Environmental 2009;88 240-248.

[52] Pereda-Ayo B, Duraiswami D, López-Fonseca R, González-Velasco JR. Influence of platinum and barium precursors on the NSR behavior of Pt-Ba/Al2O3 monoliths for lean-burn engines. Catalysis Today 2009;147 244-249.

[53] Regalbuto JR, Agashe K, Navada A, Bricker ML, Chen Q. A scientific description of Pt adsorption onto alumina. Studies in Surface Science and Catalysis 1998;118 147-156.

[54] Spieker WA, Regalbuto JR. A fundamental model of platinum impregnation onto alumina. Chemical Engineering Science 2001;56 3491-3504.

[55] Regalbuto JR, Navada A, Shadid S, Bricker ML, Chen Q. An experimental verification of the physical nature of Pt adsorption onto alumina. Journal of Catalysis 1999;184 335-348.

[56] Pereda-Ayo B, Duraiswami D, Delgado JJ, López-Fonseca R, Calvino JJ, Bernal S, González-Velasco JR. Tuning operational conditions for efficient NOx storage and reduction over a Pt-Ba/Al2O3 monolith catalyst. Applied Catalysis B: Environmental 2010;96 329-337.

[57] Pereda-Ayo B, Duraiswami D, González-Marcos JA, González-Velasco JR. Performance of NOx storage-reduction catalyst in the temperature-reductant concentration domain by response surface methodology. Chemical Engineering Journal 2011;169 58-67.

[58] Cant NW, Liu IOY, Patterson MJ. The effect of proximity between Pt and BaO on uptake, release, and reduction of NOx on storage catalysts. Journal of Catalysis 2006;243 309-317.

[59] Clayton RD, Harold MP, Balakotaiah V, Wan CZ. Pt dispersion effects during NOx storage and reduction on Pt/BaO/Al2O3 catalysts. Applied Catalysis B: Environmental 2009;90 662-676.

[60] Nova I, Castoldi L, Lietti L, Tronconi E, Forzatti P, Prinetto F, Ghiotti G. NOx adsorption study over Pt-Ba/alumina catalysts: FT-IR and pulse experiments. Journal of Catalysis 2004;222 377-388.

[61] Nova I, Castoldi L, Lietti L, Tronconi E, Forzatti P. On the dynamic behavior of "NOₓ-storage/reduction" Pt-Ba/Al₂O₃ catalyst. Catalysis Today 2002;75 431-437.

[62] Lietti L, Forzatti P, Nova I, Tronconi E. NOₓ Storage Reduction over Pt---Ba/[gamma]-Al₂O₃ Catalyst. Journal of Catalysis 2001;204 175-191.

[63] Prinetto F, Ghiotti G, Nova I, Lietti L, Tronconi E, Forzatti P. FT-IR and TPD investigation of the NOₓ storage properties of BaO/Al₂O₃ and Pt-BaO/Al₂O₃ catalysts. Journal of Physical Chemistry B 2001;105 12732-12745.

[64] Nova I, Castoldi L, Prinetto F, Dal Santo V, Lietti L, Tronconi E, Forzatti P, Ghiotti G, Psaro R, Recchia S. NOₓ adsorption study over Pt-Ba/alumina catalysts: FT-IR and reactivity study. Topics in Catalysis 2004;30-1 181-186.

[65] Forzatti P, Castoldi L, Nova I, Lietti L, Tronconi E. NOₓ removal catalysis under lean conditions. Catalysis Today 2006;117 316-320.

[66] Fridell E, Skoglundh M, Westerberg B, Johansson S, Smedler G. NOₓ storage in barium-containing catalysts. Journal of Catalysis 1999;183 196-209.

[67] Broqvist P, Gronbeck H, Fridell E, Panas I. NOₓ storage on BaO: theory and experiment. Catalysis Today 2004;96 71-78.

[68] Westerberg BR, Fridell E. A transient FTIR study of species formed during NOₓ storage in the Pt/BaO/Al₂O₃ system. Journal of Molecular Catalysis A: Chemical 2001;165 249-263.

[69] Szanyi J, Kwak JH, Hanson J, Wang CM, Szailer T, Peden CHF. Changing morphology of BaO/Al₂O₃ during NO₂ uptake and release. Journal of Physical Chemistry B 2005;109 7339-7344.

[70] Elizundia U, Lopez-Fonseca R, Landa I, Gutierrez-Ortiz MA, Gonzalez-Velasco JR. FT-IR study of NOₓ storage mechanism over Pt/BaO/Al₂O₃ catalysts. Effect of the Pt-BaO interaction. Topics in Catalysis 2007;42-43 37-41.

[71] Mahzoul H, Brilhac JF, Gilot P. Experimental and mechanistic study of NOₓ adsorption over NOₓ trap catalysts. Applied Catalysis B: Environmental 1999;20 47-55.

[72] Cant NW, Patterson MJ. The storage of nitrogen oxides on alumina-supported barium oxide. Catalysis Today 2002;73 271-278.

[73] Kikuyama S, Matsukuma I, Kikuchi R, Sasaki K, Eguchi K. A role of components in Pt-ZrO₂/Al₂O₃ as a sorbent for removal of NO and NO₂. Applied Catalysis A: General 2002;226 23-30.

[74] Rodrigues F, Juste L, Potvin C, Tempere JF, Blanchard G, Djega-Mariadassou G. NOₓ storage on barium-containing three-way catalyst in the presence of CO₂. Catalysis Letters 2001;72 59-64.

[75] Olsson L, Persson H, Fridell E, Skoglundh M, Andersson B. Kinetic study of NO oxidation and NOₓ storage on Pt/Al₂O₃ and Pt/BaO/Al₂O₃. Journal of Physical Chemistry B 2001;105 6895-6906.

[76] Olsson L, Westerberg B, Persson H, Fridell E, Skoglundh M, Andersson B. A kinetic study of oxygen adsorption/desorption and NO oxidation over Pt/Al₂O₃ catalysts. Journal of Physical Chemistry B 1999;103 10433-10439.

[77] Li XG, Meng M, Lin PY, Fu YL, Hu TD, Xie YN, Zhang J. Study on the properties and mechanisms for NO$_x$ storage over Pt/BaAl$_2$O$_4$-Al$_2$O$_3$ catalyst. Topics in Catalysis 2003;22 111-115.

[78] Cumaranatunge L, Mulla SS, Yezerets A, Currier NW, Delgass WN, Ribeiro FH. Ammonia is a hydrogen carrier in the regeneration of Pt/BaO/Al$_2$O$_3$ NO$_x$ traps with H$_2$. Journal of Catalysis 2007;246 29-34.

[79] Medhekar V, Balakotaiah V, Harold MP. TAP study of NO$_x$ storage and reduction on Pt/Al$_2$O$_3$ and Pt/Ba/Al$_2$O$_3$. Catalysis Today 2007;121 226-236.

[80] Liu ZQ, Anderson JA. Influence of reductant on the thermal stability of stored NO$_x$ in Pt/Ba/Al$_2$O$_3$ NO$_x$ storage and reduction traps. Journal of Catalysis 2004;224 18-27.

[81] Mulla SS, Chaugule SS, Yezerets A, Currier NW, Delgass WN, Ribeiro FH. Regeneration mechanism of Pt/BaO/Al$_2$O$_3$ lean NO$_x$ trap catalyst with H$_2$", Catalysis Today 2008;136 136-145.

[82] Partridge WP, Choi JS. NH$_3$ formation and utilization in regeneration of Pt/Ba/Al$_2$O$_3$ NO$_x$ storage-reduction catalyst with H$_2$. Applied Catalysis B: Environmental 2009;91 144-151.

[83] Nova I, Lietti L, Castoldi L, Tronconi E, Forzatti P. New insights in the NO$_x$ reduction mechanism with H$_2$ over Pt-Ba/gamma-Al$_2$O$_3$ lean NO$_x$ trap catalysts under near-isothermal conditions. Journal of Catalysis 2006;239 244-254.

[84] Pereda-Ayo B, González-Velasco JR, Burch R, Hardacre C, Chansai S. Regeneration mechanism of a Lean NO$_x$ Trap (LNT) catalyst in the presence of NO investigated using isotope labelling techniques. Journal of Catalysis 2012;285 177-186.

[85] Lietti L, Nova I, Forzatti P. Role of ammonia in the reduction by hydrogen of NO$_x$ stored over Pt-Ba/Al$_2$O$_3$ lean NO$_x$ trap catalysts. Journal of Catalysis 2008;257 270-282.

[86] Clayton RD, Harold MP, Balakotaiah V. NO$_x$ storage and reduction with H$_2$ on Pt/BaO/Al$_2$O$_3$ monolith: Spatio-temporal resolution of product distribution. Applied Catalysis B: Environmental 2008;84 616-630.

[87] Lindholm A, Currier NW, Fridell E, Yezerets A, Olsson L. NO$_x$ storage and reduction over Pt based catalysts with hydrogen as the reducing agent. Influence of H$_2$O and CO$_2$. Applied Catalysis B: Environmental 2007;75 78-87.

[88] Epling WS, Yezerets A, Currier NW. The effects of regeneration conditions on NO$_x$ and NH$_3$ release from NO$_x$ storage/reduction catalysts. Applied Catalysis B: Environmental 2007;74 117-129.

[89] Nova I, Castoldi L, Lietti L, Tronconi E, Forzatti P. How to control the selectivity in the reduction of NO$_x$ with H$_2$ over Pt-Ba/Al$_2$O$_3$ Lean NO$_x$ Trap catalysts. Topics in Catalysis 2007;42-43 21-25.

[90] Kabin KS, Muncrief RL, Harold MP, Li YJ. Dynamics of storage and reaction in a monolith reactor: lean NO$_x$ reduction. Chemical Engineering Science 2004;59 5319-5327.

[91] Clayton RD, Harold MP, Balakotaiah V. Performance Features of Pt/BaO Lean NO$_x$ Trap with Hydrogen as Reductant. Aiche Journal 2009;55 687-700.

Chapter 7

Optimization of Diesel Engine with Dual-Loop EGR by Using DOE Method

Jungsoo Park and Kyo Seung Lee

Additional information is available at the end of the chapter

1. Introduction

The diesel engine has advantages in terms of fuel consumption, combustion efficiency and durability. It also emits lower carbon dioxide (CO_2), carbon monoxide(CO) and hydrocarbons(HC). However, diesel engines are the major source of NOx and particulate matter emissions in urban areas. As the environmental concern increases, a reduction of NOx emission is one of the most important tasks for the automotive industry. In addition, future emission regulations require a significant reduction in both NOx and particulate matter by using EGR and aftertreatment systems (Johnson, 2011).

Exhaust gas recirculation (EGR) is an emission control technology allowing significant NOx emission reductions from light- and heavy-duty diesel engines. The key effects of EGR are lowering the flame temperature and the oxygen concentration of the working fluid in the combustion chamber (Zheng et al., 2004).

There are conventional types of EGR, high pressure loop and low pressure loop EGR. Table 1 shows advantages and drawbacks of each type of EGR.

	Advantages	Drawbacks
HPL EGR	• Lower HC and CO emissions • Fast response time	• Cooler fouling • Unstable cylinder-by-cylinder EGR distribution
LPL EGR	• High cooled EGR • Clean EGR (no fouling) • Stable cylinder-by-cylinder EGR distribution • Better Φ/EGR rate	• Corrosion of compressor wheel due to condensation water • Slow response time • HC/CO increase

Table 1. Advantages and different types of EGR

Because a HPL EGR has fast response, especially at lower speed and load, it is only applicable when the turbine upstream pressure is sufficiently higher than the boost pressure. For the LPL EGR, a positive differential pressure between the turbine outlet and the compressor inlet is generally needed. However, LPL EGR has slow response than that of HPL systems, especially at low load or speed (Yamashita et al., 2011). Facing the reinforced regulations, exhaust gas recirculation system is widely used and believed to be a very effective method for NOx and PM reductions. Furthermore, increasing needs of low temperature combustion (LTC), EGR have been issued as key technology expecting to provide heavy EGR rate and newly developed dual loop EGR system as the future of EGR types has became a common issue.

The experimental results of dual loop EGR systems were reported in Cho et al. (2008) who studied high efficiency clean combustion (HECC) engine for comparison between HPL, LPL and dual loop EGR at five operating conditions. Adachi et al.(2009) and Kobayashi et al.(2011) reported that The combination of both high boost pressure by turbocharger and a high rate of EGR are effective to reduce BSNOx and PM emissions. Especially, The EGR system using both high-pressure loop EGR and low-pressure loop EGR is also effective to reduce BSNOx and PM emissions because it maintains higher boost pressure than that of the high-pressure loop EGR system alone.

It was also reported that determination of the intake air/exhaust gas fraction by proper control logic (Wang, 2008; Yan & Wang, 2010, 2011) and turbocharger matching (Shutty, 2009) was important. However, there was more complex interaction between variables affecting the total engine system. Therefore, it is necessary to identify dominant variables at specific operating conditions to understand and provide adaptive and optimum control logic.

One of the optimization methods, design of experiment (DOE), can provide the dominant variables which have effects on dependent variables at specified operating conditions. Lee et al.(2006) studied the low pressure EGR optimization by using the DOE in a heavy-duty diesel engine for EURO 5 regulation. The dominant variables that had effects on torque, NOx and EGR rate were EGR valve opening rate, start of injection and injection mass. In their study, the optimized LPL EGR system achieved 75% NOx reduction with 6% increase of BSFC.

In this study, as one of the future EGR types, the dual loop EGR system which had combining features of high pressure loop EGR and low pressure loop EGR was developed and optimized to find the dominant parameter under frequent engine operating conditions by using a commercial engine simulation code and design of experiment (DOE). Results from the simulation are validated with experimental results.

2. Engine model

2.1. Engine specification

The engine specification used to model is summarized in Table 2. An original engine was equipped with a variable geometry turbocharger (VGT), intercooler and HPL EGR system.

Operating parameters included engine operating speed, fuel flow rate, ambient conditions, and combustion data. In addition, the length of connecting rods, distance between the piston and pin, compression ratio, and the coefficient of friction were collected and entered in GT-POWER. Data sets of valve diameters, valve timing, injection timing, duration and injection pressure were also acquired. These data were classified information of the engine manufacturer and could not be listed in detail. Engine operating conditions are summarized in Table 3.

Item	Specification
Engine volume	3 liter
Cylinder arrangement	6cyl., V- type
Bore, Stroke	84, 89mm
Compression ratio	17.3
Connecting rod length	159 mm
Wrist pin to crank offset	0.5mm
Firing order	1-3-4-2-5-6
Firing intervals	120 CA
Injection type	Common rail
EGR system	High pressure EGR system
Max. torque@rpm	240PS@3800rpm
Max. power@rpm	450N-m@1720~3500rpm

Table 2. Engine specifications

Case #	RPM	BMEP (bar)
1	732	2.17
2	1636	4.66
3	1422	3.66
4	1556	9.93
5	1909	8.80

Table 3. Engine operating conditions

The selected 5 operating conditions in the analysis were picked up from frequently operated region at emission test point given by engine manufacturer.

2.2. Engine analysis tool

Simulations were carried out by using commercial 1D code, GT-POWER, which is designed for steady-state and transient simulations and can be used for analyses of engine and powertrain control. It is based on one-dimensional gas dynamics, representing the flow and heat transfer in the piping and in the other components of an engine system. The complicated shape of intake and exhaust manifolds were converted from 3D models (by using CATIA originally) to 1D models by 3D-discretizer. Throughout the conversion, analysis of gas flow and dynamics could be faster and easier under 1D flow environment.

The combustion model was the direct-injection diesel jet (DI jet) model and it was primarily used to predict the burn rate and NOx emission simultaneously.

2.2.1. Overview of DI jet

The combustion model, DI jet, was firstly introduced by Hiroyasu known as a multi-zone DI diesel spray combustion (Hiroyasu et al., 1983). The core approach of this model is to track the fuel jet as it breaks into droplets, evaporates, mixes and burns. As such an accurate injection profile is absolutely required to achieve meaningful results. The total injected fuel is broken up into packages (also referred as zones): 5 radial and many axial slices. Each package additionally contains parcels (or subzones) for liquid fuel, unburned vapor fuel and entrained air, and burned gases.

The total mass of fuel in all of the packages will be equal to the specified injection rate (mg/stroke) divided by the specified number of nozzle holes, as DI jet will model the plume from only one nozzle hole.

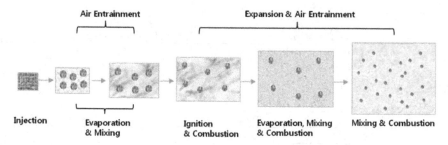

Figure 1. Air-fuel mixing process within each package

The occurring processes in the package are shown in Figure 1. The package, immediately after the fuel injection, involves many fine droplets and a small volume of air. As the package recedes from the nozzle, the air entrains into the package and fuel droplets evaporate. Therefore, the small package consists of liquids fuel, vaporized fuel and air. After a short period of time from the injection, ignition occurs in the gaseous mixture resulting in sudden expansion of the package. Thereafter, the fuel droplets evaporate, and fresh air entrains into the package. Vaporized fuel mixes with fresh air and combustion products and spray continues to burn.

2.2.2. Combustion process of DI jet

Figure 2 shows the detailed combustion process of each package in DI jet model (Hiroyasu et al, 1983). When ignition is occurred, the combustible mixture which is prepared before ignition burns in small increment of time. Combustion rate and amount of burning fuel of each package are calculated by assuming the stoichiometric condition. When the air in the package is enough for burning the vaporized fuel, there are combustion products, liquid

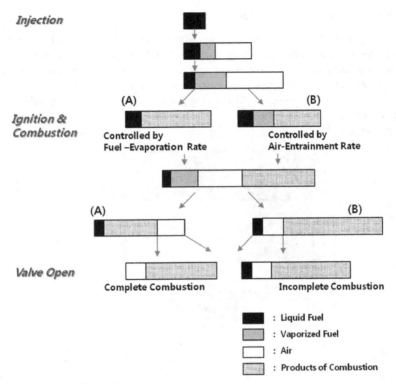

Figure 2. Schematic diagram of the mass system in a package

fuel and the remained air in the package after ignition occurs. In the next small increment of time, the fuel droplets evaporate and fresh air entrains in the package. The combustion of the next step occurs (case A in figure 2). After this step, since the stoiciometric combustion is assumed, either the vaporized fuel or the air is remained. When the air is remained, the same combustion process is repeated. But when the vaporized fuel is remained, the amount of burning fuel is controlled by the entrained air in the next step (case B in Figure 2). If ignition occurs, but the air in the package is not enough for burning the vaporized fuel, the combustion process continues under the condition shown in case B. Therefore, all the combustion processes in each package proceed under one of the conditions shown in Figure 2; Case A is evaporation rate control combustion, and B is entrainment rate control combustion. The heat release rate in the combustion chamber is calculated by summing up the heat release of each package. The pressure and average temperature in the cyinder are then calculated. Since the time histories of temperature, vaporized fuel, air and combustion products in each package are known, the equilibrium concentrations of gas compositions in the package can be calculated. The concentration of NOx is calculated by using the extended Zeldovich mechanism. More detailed governing equations can be found in Hiroyasu's studies (Hiroyasu et al, 1983).

2.3. Engine model with HPL EGR

Based on experimental data, an engine model with HPL EGR was designed. Boost pressure was matched at appropriate turbocharger speeds based on turbine and compressor maps. The injection duration at a given injection timing, injection pressure, combustion pressure and temperature were determined. Then, back pressure at the turbine downstream and EGR valve opening were determined. EGR rate, temperature and pressure drop after the EGR cooler was monitored by installing actuators and sensors. Finally, results of the simulation were compared to the experimental data. Figure 3 shows the Engine model with HPL EGR.

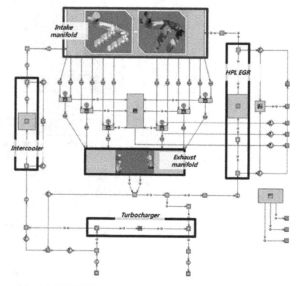

Figure 3. Engine model with HPL EGR

The percent of exhaust gas recirculation (EGR (%)) is defined as following equation.

$$EGR(\%) = \left(m_{EGR} / m_i\right) \times 100 \tag{1}$$

where [$m_i = m_a + m_f + m_{EGR}$] and m_{EGR} is the mass of EGR and m_a and m_f are the mass of air and fuel.

2.4. Engine model with dual loop EGR system and optimization

Based on the HPL model, a dual loop EGR model was designed. Comparing to the HPL EGR system, flap valve opening rate became one of the most important variables for pressure difference at P_2-P_1 in Figure 4.

First, Dual loop EGR simulation was performed under constant boost pressure. Flap valve opening at tail pipe and turbocharger RPM, which had effects on boost pressure and back pressure under dual loop EGR system, were selected as independent variables. In this case, NOx reduction rate would increase, but torque and BSFC would decrease. And the next step, optimization was performed to compensate torque loss and brake specific fuel consumption (BSFC) by modifying injection mass, start of injection (SOI) and EGR valve opening rate. Results of simulation were compared to the HPL and dual loop models in terms of torque, EGR rate, BSNOx, and BSFC.

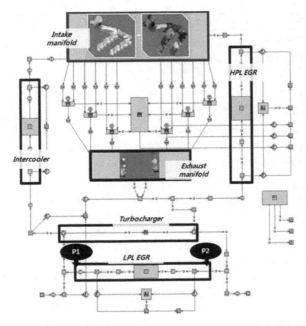

Figure 4. Engine model with dual-loop EGR

2.5. Design of experiment (DOE)

In this study, optimization based on DOE was performed.

There are main variables which have major effects on torque, BSNOx, BSFC, and EGR rate. In this study, 6 independent variables were selected such as HPL EGR valve opening diameter, LPL EGR valve opening diameter, injection mass, start of injection (SOI), flap valve opening diameter at the tail pipe and turbocharger RPM (TC RPM). Then proper ranges were set and DOE was performed based on the full factorial design.

Torque, EGR rate, BSNOx, BSFC and boost pressure were selected as response variables. The range of each independent variable was chosen based on the engine design performance.

Table 4 shows control factors and levels for optimization of the dual EGR system.

Control factor		Level 1	Level 2	Level 3
EGR valve	HPL	base	15% open	30% open
	LPL	base	15% open	30% open
Injection mass		base	+2.5%	+5%
SOI		3 CA adv.	1.5 CA adv.	base.
Flap valve		15% close	Base	15% open
TC RPM		-5000 RPM	base	+5000 RPM
Full factorial		$3^6 = 729$		

Table 4. Control factors and levels for optimization of the dual EGR system

2.5.1. Control factors

The 6 independent variables were selected, i.e. HPL EGR valve opening diameter, LPL EGR valve opening diameter, injection mass, start of injection (SOI), flap valve opening diameter at the tail pipe and turbocharger RPM (TC RPM). And their desired ranges are as follows. Base level means the values given by the experimental test.

- HPL & LPL EGR valves: If the EGR valve opens too much, it causes torque loss. In this optimization, the maximum increase of EGR valve diameter was 30% at the given operating conditions from the values under constant boost pressure.
- Injection mass: 5% increase of injection mass was selected and increasing injection mass had an effect on BSNOx. However, it normally degraded BSFC.
- SOI: In general, injection starts faster than 25-23° CA bTDC. If fuel were injected too early, imperfect combustion could degrade the engine performance. In this optimization the maximum advanced CA selected was 3 from the current value which could be within the ranges. Advanced SOI could increase torque without any other variable changes. Also, the EGR rate could increase up to 10% .
- Flap valve and TC RPM: Under dual loop EGR system, pressure difference at P2-P1 in Figure 2 was affected by interaction between flap valve and TC RPM which had a dominant effect on EGR rate and BSNOx under the dual loop EGR system.
- Especially, TC RPM was chosen to maintain boost pressure based on the turbocharger map. Positive and negative signs mean increase and decrease of rotation speed of turbocharger shaft which is driven by the exhaust flow. This change in shaft rotation speed is to optimize and maintain target boost pressure under different exhaust energy from combustion at dual-loop EGR system.

3. Results and discussion

3.1. Validation

Table 5 shows comparisons between the experiment and the simulation data in terms of injection mass and maximum cylinder pressure. There are two pilot injections and one main injection. By separating pilot injections, combustion noise, soot and NOx can be controlled.

Total injection mass and rate of main injection were given but rate of pilot injections had to be determined by matching injection duration and pressure. Figure 5 shows the torque, EGR rate, and NOx results of the simulation and the experiment, respectively. The differences of each point were within ±5% and it was proven that the simulation results had good agreement with experimental results.

Case No.	Normalized integrated injected mass (fraction)						Maximum cylinder pressure (bar)	
	Experiment			Simulation			Experiment	Simulation
	Pilot 1	Pilot 2	Main	Pilot 1	Pilot 2	Main		
1	0.134	0.134	0.732	0.120	0.130	0.750	53	53
2	0.081	0.081	0.839	0.086	0.089	0.825	58	58
3	0.102	0.102	0.797	0.097	0.122	0.781	51	51
4	0.045	0.045	0.910	0.044	0.035	0.921	75	78
5	0.047	0.047	0.905	0.049	0.045	0.906	76	79

Table 5. Comparison between the experiment and simulation data in terms of injection mass and maximum cylinder pressure

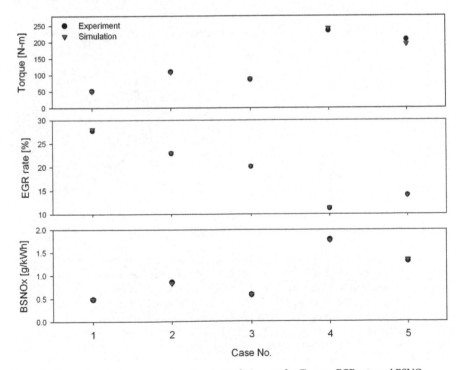

Figure 5. Comparison between experiment and simulation results: Torque, EGR rate and BSNOx

3.2. Optimization of dual-loop EGR system

Based on the DOE, response variables were determined under constant boost pressure with fixed HPL valve diameter. Then, torque and BSFC compensation were performed.

3.2.1. Optimization of dual-loop EGR system

Dual loop EGR system optimization was performed based on constant boost pressure and fixed HPL valve opening diameter to minimize torque loss. Table 6 shows the target boost pressure under dual loop EGR system which were from experiment data. Input value of HPL valve opening diameter was the same as that of the HPL model.

Case No.	1	2	3	4	5
Target boost pressure	1.03	1.23	1.11	1.43	1.56

Table 6. Target boost pressure

Figure 6 and 7 show the comparison between combustion characteristics of HPL and dual loop EGR system under constant boost pressure in case 5. Increased EGR rate caused cylinder peak pressure decrease under the dual loop EGR system. And lower peak heat release rate corresponded to lower NOx emissions.

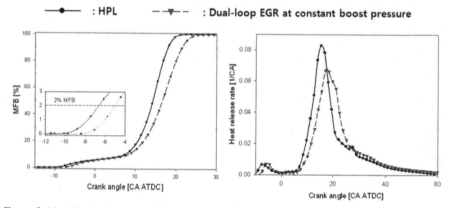

Figure 6. Mass fraction burned and heat release rate between under HPL and dual loop EGR system

Figure 8 shows the result of dual loop EGR simulation under constant boost pressure. Compared to the HPL model, 8% of torque and 8% of BSFC decreased on average. In detail, about 12 % of maximum torque loss (case 3) and about 11 % of maximum BSFC loss (case 1) occurred for the dual loop EGR. On the other hand, about 60% of NOx reduction was achieved on an average. In addition, a maximum of, 80% of NOx reduction was achieved due to the remarkable increase of the EGR rate (case 2). It seemed that the mass of the LPL EGR portion had strong effects on total NOx reduction under larger pressure difference between turbine downstream and compressor upstream. In case 3, the NOx reduction rate became lower because of smaller pressure difference at P2-P1 in Figure 4.

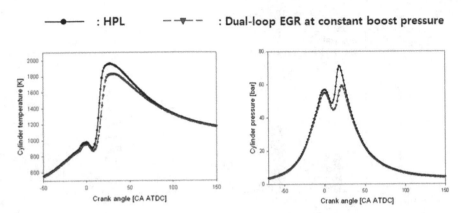

Figure 7. In-cylinder temperature and pressure trace between under HPL and dual loop EGR system

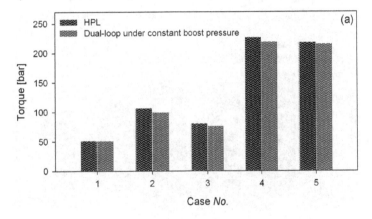

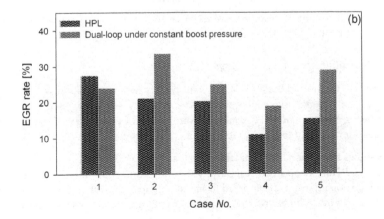

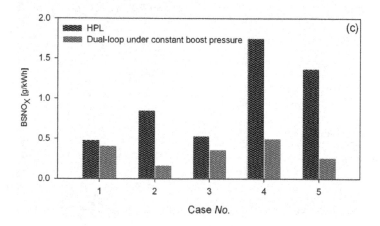

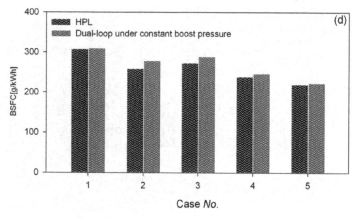

Figure 8. Simulation results comparison between HPL and dual-loop EGR under constant boost pressure; (a) Torque, (b) EGR rate, (c) BSNOx and (d) BSFC

3.2.2. *Optimization of dual-loop EGR system*

To compensate torque and BSFC under constant boost pressure condition, optimization was performed maintaining original boost pressure (bar). By advancing SOI and increasing injection mass, torque and BSFC could be compensated. Table 7 and 8 show results of optimization for constant torque and BSFC with controlled variables.

Figure 9 shows simulation results of HPL and Dual loop EGR under constant boost pressure and optimized Dual Loop EGR, respectively. 8% of torque and 5% of BSFC improvement were achieved on an average compared to the dual loop EGR system under constant boost pressure. Furthermore, higher NOx reduction efficiency appeared at each case except case 4.

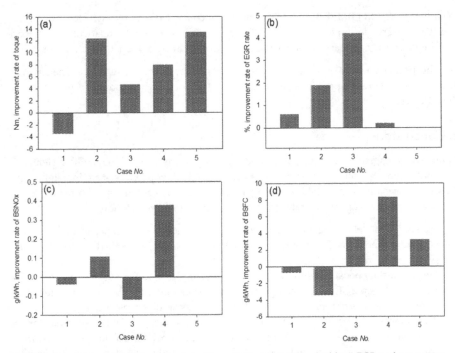

Figure 9. Results deviation of optimized dual-loop EGR compared to dual-loop EGR under constant boost pressure: (a) Torque, (b) EGR rate, (c) BSNOx and (d)BSFC

Controlled factors	Case1	Case 2	Case 3	Case 4	Case 5
HPL valve	12% open	3% open	23% open	12% open	7% open
LPL valve	12% open	9 % open	13.5% open	17% open	8% open
Injection mass	+1.5%	+3%	+2.9%	+2.5%	+1%
SOI	-	3 CA adv.	3 CA adv.	3CA adv.	1.2CA adv.
Flap valve	-	-	-	8% open	5% open
TC RPM	+2500	+2500	+5000	+1000	+2000

Table 7. Predicted values

Validation results					
Torque (N-m)	52.24	111.67	82.03	223.87	207.84
BSFC (g/kW-h)	307.2	255.27	276.25	247.01	233.30
EGR rate (%)	24.60	35.48	29.26	19.21	30.13
BSNOx (g/kW-h)	0.37	0.28	0.24	0.88	0.22

Table 8. Validation results

In case 4 (1556RPM / BMEP 9.93bar), it seemed that controlled variables which affect torque and BSFC were decoupled with the NOx reduction rate due to relatively high load conditions. It was necessary to control the variables sensitively at high load conditions.

For the optimized dual loop EGR system, 60% improvement of deNOx efficiency was achieved with increasing the EGR rates through all the cases when compared to results for the HPL system.

4. Conclusions

In this study, engine simulation was carried out to optimize the dual loop EGR system at 5 different engine operating conditions. As a result, the dual loop EGR system in a light-duty diesel engine had the potential to satisfy future emission regulations by controlling the dominant variables at given operating conditions. The details are as follows.

- An engine model for the HPL EGR was developed based on the experimental data at 5 operating conditions. The calibrated simulations showed within ±5% difference with the experimental results.
- Under constant boost pressure conditions, an average 60% NOx reduction was achieved in the dual loop EGR system compared to the results under the HPL system. However, approximately 8% of torque loss and 8% of BSFC loss occurred respectively.
- To compensate torque and fuel consumption, independent variables, such as start of injection and injection mass, were selected as additional control factors. Comparing these variables to the dual loop EGR system under constant boost pressure, approximately 8% of torque and 5% BSFC improvement were achieved except at high load conditions (case 4).

For the optimized dual loop EGR system, 60% improvement of deNOx efficiency was achieved with increasing EGR rate through all the cases compared to results for the HPL system.

Author details

Jungsoo Park
The Graduate School, Department of Mechanical Engineering, Yonsei University, Sinchon-dong, Seodaemun-gu, Seoul, Korea

Kyo Seung Lee
Department of Automotive Engineering, Gyonggi College of Science and Technology, Jeongwang-dong, Siheung-si, Gyonggi-do, Korea

5. References

Johnson, T. V. (2011), Diesel Emissions in Review, *SAE International Journal of Engines* 2011-01-0304, Vol. 4, No. 1. 143-157, doi:10.4271/2011-01-0304.

Zheng, M.; Reader, G. T. & Hawley, J. G. (2004), Diesel Engine Exhaust Gas Recirculation – A Review on Advanced and Novel Concepts, *Energy Conversion & Management*, Vol. 45, No.6, (April 2004), pp. 883-900, ISSN 0196-8904

Yamashita, A.; Ohki, H., Tomoda, T. & Nakatami, K. (2011), Development of Low Pressure Loop EGR System for Diesel Engines, *SAE Technical Paper* 2011-01-1413, ISSN 0148-7191, Detroit, Michigan, USA, April , 12-14, 2011

Cho, K.; Han, M., Wagner, R. M. & Sluder, C. S. (2008), Mixed-source EGR for Enabling High Efficiency Clean Combustion Mode in a Light-duty Diesel Engine, *SAE International Journal of Engines*, Vol. 1, No. 1, 99. 457-465, ISSN 1946-3944

Adachi, T.; Aoyagi, Y., Kobayashi, M., Murayama, T., Goto Y. & Suzuki H. (2009), Effective NOx Reduction in High Boost, Wide Range and High EGR Rate in a Heavy Duty Diesel Engine, *SAE Technical Paper* 2009-01-1438, ISSN 0148-7191, Detroit, Michigan, USA, April , 20-23, 2009

Kobayashi, M.; Aoyagi, Y., Adachi, T., Murayama, T., Hashimoto, M., Goto, Y. & Suzuki, H. (2011), Effective BSFC and NOx Reduction on Super Clean Diesel of Heavy Duty Diesel Engine by High Boosting and High EGR Rate, *SAE Technical Paper* 2011-01-0369, ISSN 0148-7191, Detroit, Michigan, USA, April , 12-14, 2011

Wang, J. (2008), Air Fraction for Multiple Combution Mode Diesel Engines with Dual-loop EGR System, *Control Engineering Practice*, Vol. 16, No. 12, (December 2008), pp. 1479-1486, ISSN 0967-0661

Yan, F. & Wang, J.(2010), In-cylinder Oxygen Mass Fraction Cycle-by-Cycle Estimation via a Lyapunov-based Observer Design, *IEEE 2010 American Control Conference*, pp. 652-657, ISBN 978-1-4244-7426-4, Baltimore, Maryland, USA, June 30- July 02, 2010

Yan, F. & Wang, J.(2011), Control of Dual Loop EGR Air-Path Systems for Advanced Combustion Diesel Engines by a Singular Perturbation Methodology, *IEEE 2011 American Control Conference*, pp. 1561-1566, ISBN 978-1-4577-0080-4, San Francisco, California, USA, June 29- July 01, 2011

Mueller, V.; Christmann, R., Muenz, S. & Gheorghiu, V. (2005), System Structure and Controller Concept for an Advanced Turbocharger/EGR System for a Turbocharged Passenger Car Diesel Engine, *SAE Paper No.* 2005013888.

Shutty, J. (2009), Control Strategy Optimization for Hybrid EGR Engines, *SAE Technical Paper* 2009-01-1451, ISSN 0148-7191, Detroit, Michigan, USA, April , 20-23, 2009

Lee, S. J.; Lee, K. S., Song, S. & Chun, K. M. (2006), Low Pressure Loop EGR System Analysis Using Simulation and Experimental Investigation in Heavy-duty Diesel Engine, *International Journal of Automotive Technology*, Vol.7, No. 6, (October 2006), pp. 659-666, ISSN 1229-9138

Park, J.; Lee, K. S., Song, S. & Chun, K. M. (2010), A Numerical Study for Light-duty Diesel Engine with Dual Loop EGR System under Frequent Engine Operating Conditions by Using DOE, *International Journal of Automotive Technology*, Vol.11, No. 5, (October 2010), pp. 617-623, ISSN 1229-9138

Hiroyasu, H.; Kadota, T. & Arai, M. (1983), Development and Use of a Spray Combustion Modeling to Predict Diesel Engine Efficiency and Pollutant Emissions: Part 1

Combustion Modeling, *Bulletin of the JSME*, Vol. 26, No. 214, (April 1983), pp.569-575, ISSN 0021-3764

Engine Control and Conditioning Monitoring Systems

Chapter 8

Model-Based Condition and State Monitoring of Large Marine Diesel Engines

Daniel Watzenig, Martin S. Sommer and Gerald Steiner

Additional information is available at the end of the chapter

1. Introduction

Although the history of diesel engines extends back to the end of the nineteenth century and in spite of the predominant position such engines now hold in various applications, they are still subject of intensive research and development. Economic pressure, safety critical aspects, compulsory onboard diagnosis as well as the reduction of emission limits lead to continuous advances in the development of combustion engines.

Condition monitoring and fault diagnosis represent a valuable set of methods designed to ensure that the engine stays in good condition during its lifecycle, [7] and [13]. Diagnosis in the context of diesel engines is not new and various approaches have been proposed in the past years, however, recent technical and computational advances and environmental legislation have stimulated the development of more efficient and robust techniques. In addition, the number of electronic components such as sensors or actuators and the complexity of engine control units (ECUs) are steadily increasing. Meanwhile, most of the software running on the main ECU is responsible for condition monitoring of sensor signals, monitoring parameter ranges, detecting short/open circuits, and verifying control deviations. However, these kinds of condition monitoring systems (CMS) are not designed to detect and clearly identify different engine failures, sensor drifts and to predict developing failures, i.e. to asses degradation of certain components right in time. Especially the reliable detection and separation of engine malfunctions is of major importance in various fields of industry in order to predict and to plan maintenance intervals.

Diesel engines usually consist of a fuel injection system, pistons, rings, liners, an inlet and exhaust system, heat exchangers, a lubrication system, bearings and an ECU. For the design of an efficient CMS it is essential to know as much as possible about the underlying thermodynamical processes and possible faults and malfunctions. This information can be seen as *a-priori* knowledge and can be used to increase the robustness of fault detection algorithms.

In the following, common diesel engine faults and fault mechanisms, and their causes are listed.

- power loss caused by misfire and blow-by.

- emission change caused by loss of compression, turbocharger malfunction, blocked fuel filter, incorrect injector timing, poor diesel fuel, incorrect fuel air ratio, air intake filter blocked, incorrect piston topping, or ECU malfunction etc.

- lubricating system fault due to incorrect oil pressure and oil deterioration

- thermal overload as a result of one or a combination of leaking injection valves, piston ring-cylinder wear or failure, eroded injector holes, too low injection pressure, high engine friction, misfire, leaking intake or exhaust manifold/valves, high coolant or lubricant temperature etc.

- leaks in the fuel injection system, lubrication system, or air intake

- wear of the piston caused by either corrosion or abrasion, or both

- noise and vibration caused by the impact of one engine part against another (mechanical noise), vibrations resulting from combustion, intake and exhaust noise

- other faults like knocking, filter faults, fuel contamination and aeration

The main challenge in engine fault detection is the ambiguity between faults and causes. Certain engine faults may be caused by a combination of causes (with different weightage). The assessment of engine states from sparse measurement data as well as a reliable assignment of failure effects and causes are an active research field. The problems relating to marine diesel engines, especially medium- and high-speed engines, are due mainly to their large size and their high operating speed. Occurring faults of marine diesel engines which are on the high seas for several months may lead to expensive holding times. On the other hand, additional sensors and measurement equipment for condition monitoring are usually undesirable since engines have to be modified to place those additional sensors. Such additional sensors are e.g. viscosity sensors to sense oil degradation as described by [12] and [1], or acoustical sensors to determine faults based on acoustical pressure and vibration signals measurements as can be found in [5], [11], and [2].

A topical review on different fault diagnosis methods for condition monitoring can be found in [7]. Both standard methods (Fourier analysis of pressure, torque, power, crankshaft speed and vibration signals) and advanced methods (neural networks, fuzzy techniques) are encountered and briefly described. [14] discuss the detection of a single fault in a statistical framework (hypotheses testing) by measuring acoustic emission energy signals and applying an independent component analysis. However, most methods usually rely on heuristic knowledge and on a data training phase as well as on the specification of threshold levels in order to assign states as faulty or non-faulty. Since the last decade, a paradigm shift from classical signal processing and feature extraction to computationally expensive model-based CMS can be observed. In contrast to classical condition monitoring, model-based methods can manage distributed and multiple correlated parameters, as described by [16] and [13]. They cover a wide variety of states since the engine behavior is described in terms of physical relationships and hence, parameters that influence certain parts of the first principles equations can be isolated or at least correlations can be determined. Three different methods to estimate the compression ratio from simulated cylinder pressure traces are presented in [8] and compared in terms of estimation accuracy and computation time. By reconstructing only one single failure based on polytropic compression and expansion of the cylinder pressure significant results have been reported. However, the detection of multiple failures from in-cylinder pressure measurements is still an open issue. Different Fuzzy-based methods also provide remarkable results for detecting only one single fault such as in [3], [4], [17], and [18].

In this work the main focus is on a robust model-based identification and separation of two common failure modes of large marine diesel engines by accurately modeling the underlying thermodynamic process. These two failures, which cause very similar changes in the cylinder pressure, are

- changes in the compression ratio primarily leading to emission and power changes
- increased blow-by mainly resulting in a loss of power.

Following a model-based approach, it is possible to identify the above mentioned failures and to clearly separate them given uncertain measurement data with low sampling rate (1° of crank angle). By measuring only cylinder pressure traces of every cylinder, the symptoms due to faults are determined, [8]. Two different approaches – ratiometric and nonlinear parameter estimation – are investigated, validated with measured data and compared to each other in terms of performance, accuracy, and robustness given sensor drift and uncertain measurements, [15].

2. The thermodynamical process model

Various approaches to model diesel engines have been proposed in literature, however, the main focus is on small-size engines that are commonly used in the automotive industry. The typical differential equations that represent the thermodynamic processes, i.e. the interrelationship between system pressure, temperature and mass can be found in [6], [9], and [10].

Since in this work, identification of blow-by and compression ratio is of primary interest, a simplified thermodynamical model capable of running in real-time is developed. Note that a list of symbols used in the following equations is given at the end of this chapter. The main reason for compression losses are referred to as damages of the piston crown during the combustion phase leading up to an increasing volume V_0 in the top dead center (TDC) of the piston. In the equation for the volume V in the cylinder the constant volume fraction V_0 is represented by the term $h_0 \cdot A$ with h_0 being the compression parameter and A the cross-sectional area of the cylinder. In the time-varying fraction of the volume equation ω denotes the instantaneous angular velocity of the crankshaft and ΔV the maximum volume deviation related to the movement of the piston. By also taking into account the ratio λ of the crank radius to the length of the connecting rod regarding to the equation of a standard crank mechanism the equation for the volume and its time derivative can be summarized as follows

$$V = h_0 \cdot A + \frac{\Delta V}{2}\left[(1 - \cos(\omega t)) + \frac{1}{\lambda} \cdot \left(1 - \sqrt{1 - \lambda^2 \sin^2(\omega t)}\right)\right] \tag{1}$$

$$\frac{dV}{dt} = \frac{\Delta V}{2} \cdot \omega \cdot \sin(\omega t)\left(1 + \frac{\lambda \cos(\omega t)}{\sqrt{1 - \lambda^2 \sin^2(\omega t)}}\right) \tag{2}$$

For the length of the connceting rod being large compared to the crank radius the terms including λ in Equation (1) and (2) can be neglected.

The time derivative of the mass fraction passing by the piston is described by

$$\frac{dm}{dt} = \tilde{k}\frac{1}{\sqrt{T}}p \tag{3}$$

with $\tilde{k}$ denoting the parameter for blow-by. For simplicity, for the *healthy state* of the cylinder it is assumed that the effect of blow-by as well as wear of the piston can be neglected. Because of the fact that blow-by is rapidly increasing when it comes to a tear-off of the oil film between piston and liner due to the loss of the sealing function of the oil the simple model of $\tilde{k}$ as a constant is not sufficient. To model this nonlinear behavior a sigmoid function described by

$$\tilde{k}(p) = \frac{\tilde{k}_{max}}{1 + e^{-a(p-b)}} \tag{4}$$

is used where b describes the pressure when 50% of the maximum blow-by is reached and a denotes the ascending slope of the sigmoid function as illustrated in Figure 1, [15]. For the

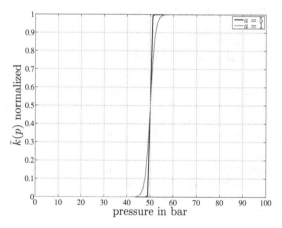

Figure 1. Sigmoid function to describe the nonlinear behavior of blow-by related to the cylinder pressure. The rapid increase of the mass passing by the cylinder results from the tear-off of the oil film at the piston crown.

complete thermodynamical description the equations for the temperature T as well as the in-cylinder pressure p represented by

$$\frac{dT}{dt} = T_{in} - \frac{p}{mc_v}\frac{dV}{dT} - \frac{kV^{\alpha_1}T^{\alpha_2}p^{\alpha_3}}{mC_v} \tag{5}$$

$$\frac{dp}{dt} = \left(RT\frac{dm}{dt} + R\frac{dT}{dt}m - p\frac{dV}{dt}\right)\frac{1}{V} \tag{6}$$

are needed containing the isochore heat capacitance C_v and the rapid increase of the temperature in the cylinder during the combustion phase T_{in} in Equation 5 and the ideal gas constant R in Equation 6. Since during our investigations only the failure parameters during the compression phase are of interest, T_{in} can be neglected. Regarding our assumption of the *healthy state* of the diesel engine with blow-by and piston wear being negligible the pair $[h_0 \ \tilde{k}] = [0.15 \ 0]$ for the compression and blow-by parameter has been identified.

NOMENCLATURE

h_0	compression parameter	m
A	cylinder cross–sectional area	m^2
V	cylinder volume	m^3
ΔV	maximum volume deviation	m^3
m	mass of the mixture	kg
T	temperatur of the mixture	°K
p	cylinder pressure	bar
R	ideal gas constant	J/(mol·K)
c_v	isochore heat capacitance	J/(kg·K)
$\tilde{k}$	blow–by parameter	
k	constant	
α_1	power of volume	
α_2	power of temperature	
α_3	power of pressure	

3. Measurement noise model

In order to obtain the goals of reliability and estimation robustness common perturbations of the cylinder pressure signal like detection uncertainties of the TDC, pressure offset p_0, and measurement noise $\mathbf{n}_k$ have to be analyzed and characterized. While the TDC offset is corrected by the manufacturer and the pressure offset can be included in the nonlinear parameter estimation approach, the task lies in finding an adequate probability density function (PDF) of the measurement noise. According to Figure 2 the measurement noise is modeled using a *Gaussian* PDF represented by

$$p(x) = \frac{1}{\sqrt{2\pi\sigma^2}} \exp\left[-\frac{(x-\mu)^2}{2\sigma^2}\right], \quad -\infty < x < \infty \tag{7}$$

where μ denotes the mean and σ^2 the variance of the random variable x. Figure 2 shows the noise data extracted from several measurements of the cylinder pressure together with the Gaussian distribution $\mathcal{N}(0, \sigma^2)$. Therefore, there exists no additional offset in the pressure signal due to measurement noise. The range of the analysis window of $[-90, -40]$ degrees to the TDC for the determination of the noise PDF was selected according to the reasonable signal to noise ratio (SNR) in this area.

4. Condition monitoring algorithms

Within this section two different model-based algorithms for condition monitoring of large diesel engine states are introduced and discussed. In the following both failures types – increased blow-by and decreased compression ratio – are denoted as errors. Thus the term compression error is used for a decreased compression ratio.

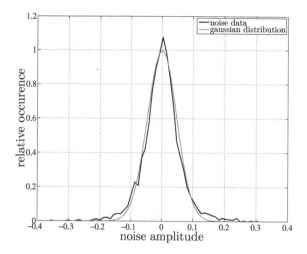

Figure 2. Histogram of the measurement noise (repeated measurements) compared to a *Gaussian* PDF (both curves are normalized by $1/\sqrt{(2\pi\sigma^2)}$).

4.1. Ratiometric approach

The main advantages of using a ratiometric approach lie in the independency of a pressure offset in the measurement data and therefore there is one disturbance variable less to be determined, the simple implementation of the method and the calculation speed. In Figure 3 all parameters for the determination of the ratiometric parameter

$$q = \frac{P_{max} - P_{min}}{P_2 - P_1} \tag{8}$$

are displayed together with two typical traces of the in-cylinder pressure of a large diesel engine. The dashed curve represents the *healthy state* whereas the solid curve reflects a cylinder state with increased compression error. The ratiometric parameter q allows to find dependencies between the error parameters h_0 and $\tilde{k}$ and the position of the analysis window $[\varepsilon_1, \varepsilon_2]$ within the compression phase. Due to the fact that blow-by has a strong nonlinear behavior causing its main influence only at hight pressures near the TDC, two analysis windows were used with the lower window being placed before and the upper window behind the inflection point of the cylinder pressure trace. To gain additional information the pressure traces in the two intervals of interest are approximated by polynomials of the form

$$P(\theta) = P_1 + a_1\theta + a_2\theta^2 + a_3\theta^3 . \tag{9}$$

As before, the *healthy state* (0% error) of the engine is described by the pair $[h_0 \ \tilde{k}] = [0.15 \ 0]$. Additionally, the maximum error (100% error) is defined by $[h_0 \ \tilde{k}] = [0.16 \ -2 \times 10^{-5}]$. The procedure is described by a case study with simulated data with 70% compression and 10% blow-by error. (see Figures 4 to 6).

The first parameter to be evaluated is the ratiometric parameter q. As can be seen in Figure 4, q alone is not sufficient to distinguish between the two failure modes. Therefore the additional

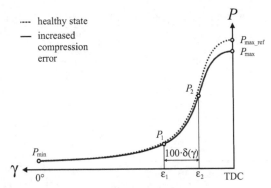

Figure 3. Typical cylinder pressure traces representing a *healthy state* (dashed) and a cylinder with increased compression error (solid). The analysis window $\delta(\gamma)$ is applied within the well-defined compression phase in order to avoid the influence of combustion effects as well as measurement noise at low signal levels.

parameters slope a_1 and curvature a_2 need to be evaluated for failure separation. As can be seen in Figure 5 the compression failure is overestimated by $\sim 10\%$ and a separation of the two failure modes is still not possible, respectively. The evaluation of the curvature information a_2 depicted in Figure 6 allows the distinction between compression and blow-by error but the compression error is still overestimated. Because of the small values of a_2 the disturbance of the curve by measurement noise with $\sigma = 0.047354$ bar becomes visible. In this sensitivity to measurement noise lies the main drawback for this method. Therefore, for the utilization of the ratiometric principle on real measurement data, Equation 8 for the calculation of the ratiometric parameter has to be modified to

$$q_{\mathrm{mod}}(\theta) = \frac{P_{\mathrm{defect}}}{P_{\mathrm{healthy}}} \tag{10}$$

with P_{defect} representing a cylinder with either blow-by or compression failure.

In Figure 7 the different curvature of q_{mod} can be determined. As can be seen the greatest differences occur at crank angles close to the TDC which are partly outside of the observation window limited by the upper bound of $-8°$ to the TDC.

In Table 1 the coefficients according to Equation 9 are summarized for three cylinders with known failure sources of two different engines excluding the pressure offset. Here the different signs for the coefficients a_1 and a_3 for blow-by and compression failure have to be noted.

occurred failures	a_1	a_2	a_3
increased blow-by	-3.99×10^{-4}	-3.57×10^{-6}	-1.09×10^{-8}
changed compression ratio	5.38×10^{-3}	-4.99×10^{-4}	1.62×10^{-5}
	3.32×10^{-3}	-3.98×10^{-4}	1.29×10^{-5}

Table 1. Coefficients of the fitting polynomial.

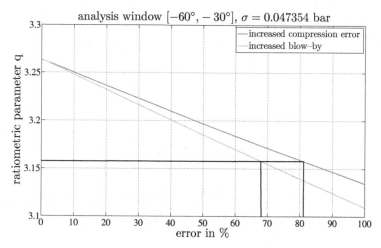

Figure 4. Ratiometric parameter q for the lower analysis window $[-60°, -30°]$ to TDC ($q = 3.1587$). The lines for compression and blow-by error are proceeding too close for a failure separation.

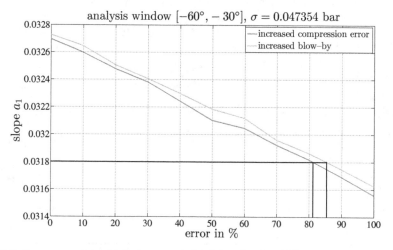

Figure 5. Slope parameter a_1 for the lower analysis window $[-60°, -30°]$ to TDC ($a_1 = 3.180 \times 10^{-2}$). The lines for compression and blow-by error are still too close together for a separation of the failure modes.

Due to the limitations and the fact that the ratiometric approach only allows a qualitative statement led to the development of a nonlinear parameter estimation approach.

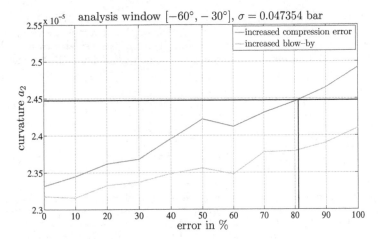

Figure 6. Curvature parameter a_2 for the lower analysis window $[-60°, -30°]$ to TDC ($a_2 = 2.447 \times 10^{-5}$). Because there is only one failure mode in the allowed range a_2 allows the separation between blow-by and compression error.

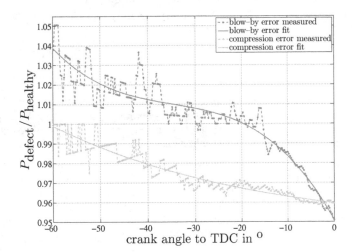

Figure 7. Comparison of the ratio $P_{\text{defect}}/P_{\text{healthy}}$ for an engine showing blow-by error and compression error respectively showing different curvature in their slopes especially in the interval $[-40°, 0°]$.

4.2. Nonlinear parameter estimation

The proposed approach aims at finding a parameter vector $\theta = [h_0 \; \tilde{k} \; p_0]^T$ which is comprised of the compression ratio h_0, the blow-by parameter $\tilde{k}$ and the pressure offset p_0 by minimizing the L_2-norm of the error $\|e\|_2^2 \to$ min between measured data and computed cylinder pressure in a nonlinear least squares sense for each cycle. The block diagram is shown in Figure 8. The disturbance of the data $\tilde{\mathbf{y}}_k$ due to measurement noise $\mathbf{n}_k$ is considered by an additional summation node with the output $\mathbf{y}_k$ representing the corrupted data. The thermodynamic

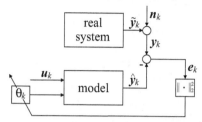

Figure 8. Block diagram of the parameter identification of large diesel engines. The index k indicates the iterative nature of the optimization procedure. By minimizing the residual error between measured and calculated cylinder pressure, the optimal parameter configuration for blow-by and compression ratio is found. Based on the *a priori* known limits of the parameters, the engine state can be assessed and monitored.

model is calibrated for a measured *healthy state* prior to the parameter identification by adapting the parameter vector $\mathbf{u} = [\alpha_1 \ \alpha_2 \ \alpha_3 \ R \ c_v \ k]^T$. The parameter identification problem consists of finding the set of parameters $\theta \in \mathbb{R}^n$ that minimizes the target function $f(\theta)$ at a single point. The inequality constraints simultaneously have to be satisfied at this single point where both the target and constraint functions depend on the parameter vector. The objective is to find a parameter configuration that satisfies

$$\min_{\theta} f(\theta) = \min_{\theta} \|\mathbf{y}_k - \hat{\mathbf{y}}_k(\theta)\|_2^2$$
$$\text{s.t.} \quad \mathbf{b}_l \le \theta \le \mathbf{b}_u \tag{11}$$

where $\mathbf{y}_k$ denotes the measured cylinder pressure and $\hat{\mathbf{y}}_k(\theta)$ represents the estimated cylinder pressure based on the thermodynamic model. The bounds $\mathbf{b}_l$ and $\mathbf{b}_u$ are the lower and the upper bound for the unknown parameter vector, i.e. the imposed constraints on the parameters to be reconstructed from measured data.

In order to mask out undesired effects of the starting combustion close to the TDC and the low SNR at small cylinder pressures, a window function $\delta(\gamma)$ is applied to the measured cylinder pressure $\mathbf{y}_k$ according to Equations (12) and (13). The proposed rectangular window is mainly restricted to the compression phase. If the entire signal $\mathbf{y}_k$ is provided to the parameter identification problem, a robust detection and identification of blow-by and compression ratio failures is impossible since various other effects influence the cylinder pressure during combustion.

$$\mathbf{z}_k = \delta(\gamma)\mathbf{y}_k \tag{12}$$
$$\delta(\gamma) = \begin{cases} 1 & \text{if } \varepsilon_1 \le \gamma \le \varepsilon_2 \\ 0 & \text{else} \end{cases} \tag{13}$$

where $\gamma \in [0 \ 360]$ denotes the crank angle in degrees. The lower and upper bound for the analysis window are given by $\varepsilon_1 = \text{TDC} - 120°$ and $\varepsilon_2 = \text{TDC} - 8°$. The model-based estimation of θ is based on the windowed signal $\mathbf{z}_k$ by solving the constrained nonlinear optimization problem (11). Signal parts with low signal magnitude as well as the signal part that corresponds to the combustion phase depicted in Figure 3 are cut off for the estimation procedure. The dashed curve representing the *healthy state* is used to calibrate the thermodynamical model by adapting the model parameter vector **u**.

5. Condition monitoring results

In the following, results for two measured data sets of different engines containing single blow-by and single compression ratio failure are presented. For the evaluation of the source of defect the engines were disassembled by the manufacturer. The reason for lower compression ratios was identified as burn-off of the piston crown whereas increased blow-by occurred due to defects of the crankcase cover gasket. The model limits for the parameters to be estimated are $h_0 = [0.15\ 0.16]$ and $\tilde{k} = [0\ -2 \times 10^{-5}]$ corresponding to $[0\%\ 100\%]$ of failure. The main objective is to identify and to quantify the occurring failures. Figure 9 illustrates the pressure traces of a five cylinder diesel engine, respectively.

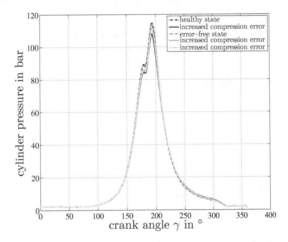

Figure 9. Measured cylinder pressure traces representing a *healthy state* and cylinders with increased compression error of one specific engine.

In both cases the sources of defect were known. Because blow-by errors often lead to severe damages of the engine most of the time the crankcase cover gaskets are replaced before the error occurs and therefore there exist only a few data sets where blow-by is documented.

Figure 10 shows such a case for one cylinder of a seven cylinder diesel engine. As can be seen the single pressure traces are close together up to the TDC. As the observation window is limited by $-8°$ to the TDC, the area with the greatest change in the cylinder pressure cannot be used which makes the detection and separation of the interesting failures a challenging task. For quantification the model limits for the parameters to be estimated are again $h_0 = [0.15\ 0.16]$ for compression and $\tilde{k} = [0\ -2 \times 10^{-5}]$ for blow-by corresponding to $[0\%\ 100\%]$ of failure. Table 2 summarizes the results of the estimated parameter vector θ with varying upper bound ε_2 of the analysis window. The first block indicates increased blow-by given the

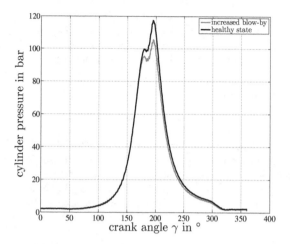

Figure 10. Measured cylinder pressure representing a *healthy state* and one cylinder with increased blow-by of a different engine. The sources of defect were in both cases documented by the manufacturer after disassembly of the machine.

desired compression ratio while in the second block a changed compression ratio is clearly identified.

occurred failures	$\gamma-$TDC	h_0	$\tilde{k}$
increased blow-by	$-8°$	0.150	-1.84×10^{-5}
	$-9°$	0.150	-1.67×10^{-5}
	$-10°$	0.150	-1.99×10^{-5}
changed compression ratio	$-8°$	**0.154**	-3.01×10^{-15}
	$-9°$	**0.154**	-2.72×10^{-15}
	$-10°$	**0.154**	-1.72×10^{-15}

Table 2. Results for estimated blow-by and compression ratio.

The blow-by estimate remains very small denoting that blow-by has not increased. In addition, the nonlinear approach exhibits a robust parameter estimation behavior given uncertainties in TDC within a certain range.

6. Conclusions

This book chapter addresses different methods for robust detection of increased blow-by and compression faults from measured cylinder pressure traces of large marine diesel engines. By modeling the underlying thermodynamic process, including prior knowledge about the system, and characterizing the measurement noise, faults can be detected and isolated from each other even in the presence of sensor drift.

The ratiometric approach allows only qualitative statements and can not clearly distinguish between the two failure modes blow-by and compression losses. The main drawbacks of the algorithm are its sensitivity to measurement noise and the fact that crank angles close to the TDC are required to see a proper curvature in the ratio of the pressure. On the other hand, the method is very fast due to its simplicity and independent to a pressure offset in the measurement signal. In contrast, the nonlinear parameter estimation methodology features higher accuracy in estimation results and allows to distinguish between certain types of faults, however, introducing a greater modeling effort and computational costs.

The applicability of the model-based approaches is verified by measurement data given information about the sources of defect of the engine. Due to the low sampling interval of 1° of the crank angle the condition monitoring system (CMS) exhibits real-time performance. The robustness is investigated by analyzing the statistics of the estimated parameters of blow-by and compression ratio. Furthermore, the influence of the upper limit of the analysis window close to the TDC is examined. The detection of these failures can be used in order to predict maintenance intervals. Based on cylinder pressure traces the proposed methods feature the applicability to other domains including large trucks, rail vehicles, and stationary power stations.

Author details

Daniel Watzenig
Graz University of Technology and Virtual Vehicle Research Center, Austria

Martin S. Sommer
Graz University of Technology, Austria

Gerald Steiner
Graz University of Technology, Austria

7. References

[1] Agoston, A., Ötsch, C. & Jakoby, B. [2005]. Viscosity sensors for engine oil condition monitoring–application and interpretation of results, *Sensors and Actuators A: Physical* Vol. 121: 327–332.

[2] Barelli, L., Bidini, G., Buratti, C. & Mariani, R. [2009]. Diagnosis of internal combustion engine through vibration and acoustic pressure non-intrusive measurements, *Applied Thermal Engineering* Vol. 34: 1707–1713.

[3] Çelik, M. & Bayir, R. [2007]. Fault detection in internal combustion engines using fuzzy logic, *Proceedings of the Institution of Mechanical Engineers, Part D: Journal of Automobile Engineering* Vol. 221: 579–587.

[4] Cruz-Peragon, F., Jimenez-Espadafor, F., Palomar, J. & Dorado, M. [2008]. Combustion faults diagnosis in internal combustion engines using angular speed measurements and artificial neural networks, *Energy & Fuels* Vol. 22: 2972–2980.

[5] Geng, Z. & Chen, J. [2005]. Investigation into piston-slap-induced vibration for engine condition simulation and monitoring, *Journal of Sound and Vibration* Vol. 282: 735–751.

[6] Heywood, J. B. [1988]. *Internal Combustion Engine Fundamentals*, McGraw-Hill.

[7] Jones, N. B. & Li, Y.-H. [2000]. A review of condition monitoring and fault diagnosis for diesel engines, *Tribotest Journal* Vol. 6(No. 3): 267–291.

[8] Klein, M. & Eriksson, L. [2006]. Methods for cylinder pressure based compression ratio estimation, 2006-01-0185, *SAE Technical Paper Series, SAE World Congress, Detroit, USA, April 3-6*.

[9] Kouremenos, D. A. & Hountalas, D. T. [1997]. Diagnosis and condition monitoring of medium-speed marine diesel engines, *Tribotest Journal* Vol. 4(No. 1): 63–91.

[10] Liu, H.-Q., Chalhoub, N. G. & N.Henein [2001]. Simulation of a single cylinder diesel engine under cold start conditions using simulink, *Journal of Engineering for Gas Turbines and Power* Vol. 123: 117–124.

[11] Liu, S., Gu, F. & Ball, A. [2006]. Detection of engine valve faults by vibration signals measured on the cylinder head, *Proceedings of the Institution of Mechanical Engineers, Part D: Journal of Automobile Engineering* Vol. 220: 379–386.

[12] Macián, V., Tormos, B., Olmeda, P. & Montoro, L. [2003]. Analytical approach to wear rate determination for internal combustion engine condition monitoring based on oil analysis, *Tribology International* Vol. 36: 771–776.

[13] McDowell, N., Wang, X., Kruger, U., McCullough, G. & Irwin, G. [2006]. Fault diagnosis for internal combustion engines, *Automation Technology in Practice* Vol. 3: 19–26.

[14] Pontoppidan, N. H., Sigurdsson, S. & Larsen, J. [2005]. Condition monitoring with mean field independent components analysis, *Mechanical Systems and Signal Processing* Vol. 19: 1337–1347.

[15] Watzenig, D., Steiner, G. & Sommer, M. S. [2008]. Robust estimation of blow-by and compression ratio for large diesel engines based on cylinder pressure traces, *Instrumentation and Measurement Technology Conference, IMTC 2008*, pp. 974–978.

[16] Woud, J. K. & Boot, P. [1993]. Diesel engine condition monitoring and fault diagnosis based on process models, *20th International Congress on Combustion Engines, London*.

[17] Wu, J.-D., Chiang, P.-H., Chang, Y.-W. & j. Shiao, Y. [2008]. An expert system for fault diagnosis in internal combustion engines using probability neural networks, *Expert Systems with Applications* Vol. 34: 2704–2713.

[18] Yong, X., Guiyou, H., Chunrong, S., Zhibing, N. & Wu, Z. [2010]. Reconstruction of cylinder pressure of i.c. engine based neural networks, *2010 First International Conference on Pervasive Computing, Signal Processing and Applications*.

Chapter 9

Design and Field Tests of a Digital Control System to Damping Electromechanical Oscillations Between Large Diesel Generators

Fabrício Gonzalez Nogueira, José Adolfo da Silva Sena, Anderson Roberto Barbosa de Moraes, Maria da Conceição Pereira Fonseca, Walter Barra Junior, Carlos Tavares da Costa Junior, José Augusto Lima Barreiros, Benedito das Graças Duarte Rodrigues and Pedro Wenilton Barbosa Duarte

Additional information is available at the end of the chapter

1. Introduction

Electromechanical oscillations are natural phenomena in power systems having two or more synchronous generating units operating interconnected. These oscillations are undesirable because they can severely limit the power transfer between interconnected generating areas, due to reduced stability margins, as well as may decrease lifetime expectancy of system machines. If these electromechanical oscillations are not satisfactorily damped, they may, under some operating conditions, even increase (in amplitude), causing shutdown (tripping) of one or more interconnected generating units. As the oscillations are related to the physical nature of the electrical power system component's interactions, they cannot be avoided. However, by using efficient automatic control techniques, the electromechanical oscillations can be sufficiently attenuated in order assure a safe system operation, for all allowed operating conditions [1, 2].

Among the devices utilized to deal with electromechanical oscillations, the most common are the power systems stabilizers (PSS). As can be seen in Figure 1, dashed box (a), PSS devices usually actuates through the automatic voltage regulator (AVR) in order to increase the damping of poorly damped oscillations modes. The improvement of the damping is obtained through a torque component proportional to the machine speed deviation [1, 2].

An alternative technique, which has been investigated by several authors [3-5], is the application of a damping controller through the speed governor system (see dashed box (b), in Figure 1). This is the approach followed by this work. It is important to remark that this technique is recommended only for generators systems having fast response actuators, such as diesel engines. The installation of a damping controller via the speed regulation system may be advantageous because theoretical studies show that there is a weak coupling between this control loop and excitation system controllers in other machines of the interconnected power system [3, 4].

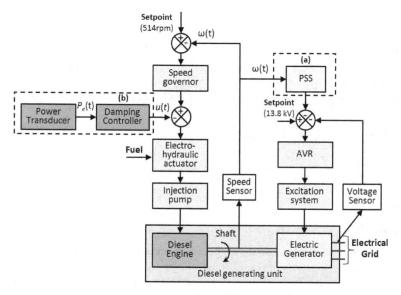

Figure 1. Comparison between a conventional PSS (a) and a damping controller actuating through the speed governor system (b).

A complete real world example of a control development for safe operation of a power plant has been presented in this chapter. The chapter details the design, implementation and field tests of a digital damping controller applied to a diesel generating unit, at Santana thermoelectric power plant (located in north of Brazil), addressing the system identification and the implementation of the damping controller using digital control techniques. The experimental tests results are presented and discussed.

2. System description

2.1. Diesel generating units at Santana Power Plant

Santana Power Plant has a set of seven generating units, including four identical 18-MVA Wärtsillä diesel generating units (Figure 2). Due to economic operation constraints, it is advisable to preferentially operating the diesel engines because these machines have the lowest specific fuel consumption among the power plant generating units.

Figure 2. Diesel generating units, at Santana Thermoelectric Power Plant.

The diesel generating units are driven by 18V46 Wärtsillä four-stroke diesel engines, which have 18 cylinders in V-form (V-angle of 45°), as shown in Figure 3. The engine presents a turbocharged and intercooled design, along with direct injection subsystem.

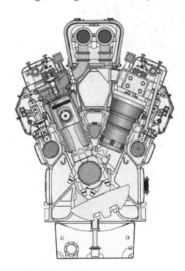

Figure 3. Cross section of the V-engine [6].

A simplified block diagram of the diesel generating units is shown in Figure 4. The turbocharging is performed by turbocompressors, which are driven by the exhaust gases from the combustion chamber. The turbine drives the compressor subsystem which in turn draws air from the environment increasing the air pressure. The compressed air is then cooled and, subsequently feedback into the combustion chamber, as illustrated in Figure 4. In order to keep the rotor speed at nominal value, during the system operation, a speed governor controls the loading of the diesel engine by actuating on the electro-hydraulic actuator position (see Figure 4). Therefore, more or less fuel is injected in order to increase or decrease the mechanical power demanded by the electrical generator and its load.

The damping controller proposed in this chapter actuates in the output of the speed governor, modulating the electro-hydraulic actuator position according to the observed electromechanical oscillation on the measured electric power. Table 1 shows the main technical characteristics of the 18V46 diesel engine:

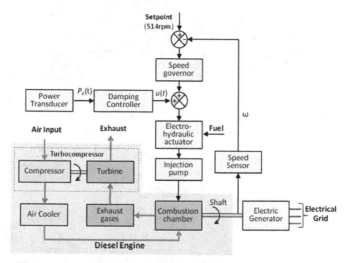

Figure 4. Simplified schematic of the diesel generating unit equipped with the damping controller.

Main technical data of engine		
Parameter	Unit	Value
Engine output	kW	17,550
Generator output	kW	17,025
Specific consumption (Diesel oil)	Liter/kWh	0.22
Cylinder bore	mm	460
Piston stroke	mm	580
Cylinder output	kW/cyl.	975
Engine speed	rpm	514
Piston speed	m/s	9.9
Mean effective pressure	bar	23.6
Firing pressure	bar	180
Charge air pressure	bar	3.1
Weight (Engine + generator)	Ton	290

Table 1. Main technical characteristics of the 18V46 diesel engine.

Depending on fuel availability, the 18V46 Wärtsillä diesel engines can be feed by one of the following fuel types: light fuel oil (light oil-diesel), heavy fuel oil, and natural gas. In Santana thermoelectric power plant, the light oil-diesel is the preferential choice. The light oil-diesel characteristics are presented in Table 2.

Light oil-diesel characteristics		
Parameter	Unit	Value
Viscosity, max	cSt a 50°C	11
Density, max	g/ml	0.92
Sulphur, max	% mass	2
Vanadium, max	mg/kg	100
Ash, max	% mass	0.05
Water, max	% vol	0.3
Pour point	ºC	6

Table 2. Light oil-diesel characteristics.

The fuel injection system is composed of injection pumps, high pressure pipes, and injection valves. The nozzle is located at the top center of the cylinder head. The pressurized fuel in the low-pressure oil line (8.0 to 11.0 bar) has its flow controlled by the electro-hydraulic actuator, which is driven by the output signal of the speed governor (Figure 5). The fuel injected into the high-pressure combustion chamber (180 bar) is atomized and the combustion occurs after the compression cycle.

Figure 5. Electro-hydraulic actuator subsystem.

2.2. Low-damped electromechanical oscillations

Based on a priori information, provided by utility technical reports [7], a dominant 2.5 Hz intra-plant electromechanical oscillation mode was identified having a much reduced damping, which was observed from the measured machine terminal electrical power signal (Figure 6), when was applied a step in the electro-hydraulic actuator of the generating unit 2. As can be seen, diesel generating units 2 and 3 (G2 and G3) oscillated on phase opposition, which is interpreted as an indicative of intra-plant oscillation mode [2]. These tests were performed in order to investigate the reasons for the frequent machine shutdown (trip) due to actuating of torsional mode protection. Based on this a priori information, the 2.5 Hz intra-plant oscillation mode was then chosen as the target oscillation mode for which the digital controller presented in this chapter was designed.

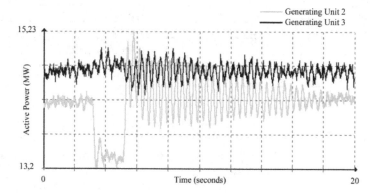

Figure 6. G2 and G3 response to a step variation applied to the G2 electro-hydraulic actuator [7, 8].

3. Digital damping controller

In order to perform the field tests, the damping controller was implemented on an embedded system, composed by the main blocks: input conditioning system, digital controller, actuation system, local HMI, and communication interface (Figure 7).

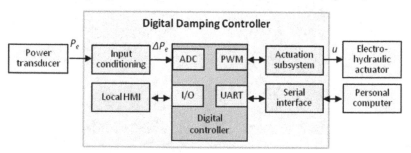

Figure 7. Damping controller blocks.

In order to obtain a signal having enough information about the dominant power oscillation, it was necessary to pre-processing the measurement data. To that end, the active power signal (P_e) has been chosen as the feedback signal to the damping controller. This choice was based on field tests, which indicated that the signal/noise ratio can be improved. Therefore, the P_e signal was applied in the conditioning module, where it was filtered by a first-order low-pass filter in order to attenuate noise and, after that, the signal was processed by a first-order high-pass filter (washout) in order to eliminate the DC component.

The digital processing module of the damping controller is based on a digital signal controller (DSC). This device incorporates features of microcontrollers (variety of internal peripherals) and the DSP's (specific support for digital signal processing). The analog signal ΔP_e is converted to digital through an internal 12-bits analog-to-digital converter (ADC). The firmware of the proposed damping controller control law was embedded in the digital signal controller and its control law was programmed by using C language.

The output of the digital controller is generated through an internal pulse width modulation (PWM) module of the DSC. The action of the damping controller in the electro-hydraulic actuator of the diesel engine is achieved by means of an actuator subsystem, which modulates the output current of the speed regulator. This electronic sub-system is implemented as an array of power transistors, which composes a current mirror scheme, which is commanded by the PWM signal.

The communication between the embedded system and other devices is performed through a RS-232 serial interface. Through this communication link, the embedded system is able to transmit the collected data to a personal computer (PC), to perform analyzing and control.

The developed embedded system was designed to operate in four different operational modes, namely: (i) "step response", (ii) "identification", (iii) "control", and (iv) "configuration mode". When in the first mode, the equipment applies a step variation to the electro-hydraulic actuator, collect the plant response and send the collected data to an auxiliary microcomputer PC, for a more complex data analysis. When operating in identification mode, the damping controller may generate a pseudo-random binary sequence (PRBS) and uses this signal test to modulate the electro-hydraulic actuator position of the diesel engine. Using a non-recursive least mean squares algorithm, pairs of input and output data are used to obtain plant estimated parametric models, which are used on the controller design. When in "control mode", the damping controller acquires the signal of active power deviation and processes the damping control law, generating a control signal that is applied in the actuator of the fuel valve of the diesel motor, in order to damp the electromechanical oscillations. Finally, in "configuration mode", the user can set up configuration parameters of the controller through a keyboard and a menu on a LCD display (local HMI).

4. Identification of linear models

4.1. Model structure

In order to design the digital damping controller, it is necessary to estimate a mathematical model that represents the system dynamics at a specific operation condition. This modeling step can be performed using identification techniques, establishing a system dynamical model from measured input and output data, which are respectively, the current of command of the fuel valve and the signal proportional to the active power of the generator. The identified model captures the relevant information about the plant dynamic, for feedback control objectives.

A dynamic system model can be represented in several different ways, particularly in time and frequency domains. When the objective is to obtain a dynamic linear model around an operation point using sampled data, a discrete linear parametric model can be used such as an autoregressive with exogenous inputs (ARX) model. The ARX model can be represented in the discrete time domain by [9, 10]:

$$A(q^{-1})y(k) = B(q^{-1})u(k) + v(k) \tag{1}$$

Or by the transfer function:

$$G_p(q^{-1}) = B(q^{-1})/A(q^{-1}) \tag{2}$$

where q^{-1} is the discrete-time delay operator, $y(k)$ and $u(k)$ are the sequences of input and output data, respectively, and $v(k)$ is assumed to be a white noise, a signal having a flat power spectral density, which is a very useful property for analysis and system identification [9, 10]. $B(q^{-1})$ and $A(q^{-1})$ are discrete-time polynomials in the form:

$$B(q^{-1}) = b_1 q^{-1} + b_2 q^{-2} + \cdots + b_{nb} q^{-nb} \tag{3}$$

$$A(q^{-1}) = 1 + a_1 q^{-1} + a_2 q^{-2} + \cdots + a_{na} q^{-na} \tag{4}$$

Coefficients $b_1, ..., b_{nb}$ and $a_1, ..., a_{na}$ are the model parameters, and n_b and n_a are integer numbers used to define the model order.

4.2. Estimation algorithm

The input-output (I/O) representation (1) can be put into a linear regression form as follows. Let us define a $n = n_a + n_b$ vector which contains all the coefficients to be identified:

$$\theta^T = [a_1\ a_2\ \cdots\ a_{na}\ b_1\ b_2\ \cdots\ b_{nb}] \tag{5}$$

We also define the extended regressor, which will be made up of past I/O data:

$$\phi^T(k) = [-y(k-1)\ -y(k-2)\ \cdots -y(k-n_a)\ \ u(k-1)\ u(k-2)\ \cdots\ u(k-n_b)] \tag{6}$$

The model output $\hat{y}(k)$ can be calculated through the product $\hat{\theta}^T(k)\phi(k)$. Therefore, the estimating error $\varepsilon(k)$ is the difference between the measured system output $y(k)$ and the estimated output $\hat{y}(k)$, i.e., $\varepsilon(k) = y(k) - \hat{\theta}^T(k)\phi(k)$.

The actualization of the model parameters $\hat{\theta}$ can be performed through an iterative algorithm, such as the least mean square method (LMS):

$$\varepsilon(k) = y(k) - \hat{\theta}^T(k)\phi(k) \tag{7}$$

$$\hat{\theta}(k+1) = \hat{\theta}(k) + \gamma \varepsilon(k)\phi(k) \tag{8}$$

where the parameter γ is the step size.

4.3. Persistency of excitation and data acquisition

Before data acquisition, a previous system study is necessary to find the best way to obtain dynamic information of the plant. Analyzing system modes frequencies, an appropriate excitation test signal can be designed, which will excite the plant in a range of desired frequencies. This operation results in a better capture of dynamic information of the system, thus improving the dynamical model estimation.

The system to be identified has its dominant modes in a characteristic range of frequencies. Therefore, to excite the plant in an appropriate manner, the input signal must be designed to have an approximate uniform power spectrum in the dominant range. An exciting signal satisfying this property is the pseudo-random binary sequence (PRBS), which is a binary signal that can be generated by using digital techniques [10]. This can be done by using a feedback shift register having a length of N cells and a sample generation interval T_b. From knowledge of the minimum and maximum values f_{min} and f_{max} for the desired range of frequency excitation, the values of N and T_b can be calculated by using equations (9) and (10) [9, 10, 11]:

$$f_{max} = \frac{0{,}44}{T_b} \tag{9}$$

$$f_{min} = \frac{1}{(2^N - 1)T_b} \tag{10}$$

5. Design of the controller to damp electromechanical oscillations

5.1. Pole shifting technique

The damping controller goal is to increase the dominant mode damping, without changing significantly the natural frequency (ω_n) of this mode. In order to perform this task, the pole shifting technique was utilized to obtain a controller able to provide a stable closed loop system and performance characteristics as specified in accordance with the designer requirements. This method is a particular case of the general pole placement method.

In this technique, the open-loop dominant poles must be radially shifted to a new position toward the origin of the unitary circle in the z-plane. The amount of the radial displacement is specified by a contraction factor α, according to the desired degree of damping for the closed loop system [12, 13]. Therefore, the designer first specifies a desired value, ξ_d, for the damping of the electromechanical mode and then calculates the value of the shifting factor α, by using:

$$\alpha = e^{-(\xi_d - \xi)\omega_n T_s} \tag{11}$$

where $0 \leq \alpha \leq 1$, T_s is the controller sampling period, ξ is the natural damping (system without damping controller).

The pole shifting method is based on the search of polynomials $R(q^{-1})$ and $S(q^{-1})$, which satisfy the polynomial equation (12), known as Diophantine equation:

$$A(q^{-1})S(q^{-1}) + B(q^{-1})R(q^{-1}) = A_{cl}(q^{-1}) \tag{12}$$

Where the polynomials $A(q^{-1})$ and $B(q^{-1})$ are both known for the designer (model estimated of the plant), $A_{cl}(q^{-1})$ is a polynomial with the desired closed-loop poles, and $R(q^{-1})$ and $S(q^{-1})$ are the polynomials with the parameters of the controller to be calculated:

$$R(q^{-1}) = r_0 + r_1 q^{-1} + r_2 q^{-2} + \cdots + r_{nr} q^{-nr} \tag{13}$$

$$S(q^{-1}) = 1 + s_1 q^{-1} + s_2 q^{-2} + \cdots + s_{ns} q^{-ns} \tag{14}$$

Assuming that $n_a = n_b = n$ and $n_r = n_s = n - 1$ (minimum order controller), and matching the coefficients of the same power in q^{-1}, the coefficients of the control law can be obtained directly by the solution of the following linear equation system:

$$\begin{bmatrix} 1 & 0 & . & 0 & b_1 & 0 & . & 0 \\ a_1 & 1 & . & 0 & b_2 & b_1 & . & 0 \\ . & a_1 & . & . & . & b_2 & . & . \\ a_{n_a} & . & . & 1 & b_{n_b} & . & . & b_1 \\ 0 & a_{n_a} & . & a_1 & 0 & b_{n_b} & . & b_2 \\ . & 0 & . & 0 & . & 0 & . & . \\ . & . & . & . & . & . & . & . \\ 0 & 0 & . & a_{n_a} & 0 & 0 & . & b_{n_b} \end{bmatrix} \begin{bmatrix} s_1 \\ . \\ . \\ s_{n_s} \\ r_0 \\ . \\ . \\ r_{n_r} \end{bmatrix} = \begin{bmatrix} (\alpha - 1)a_1 \\ (\alpha^2 - 1)a_2 \\ . \\ . \\ (\alpha^{n_a} - 1)a_{n_a} \\ 0 \\ . \\ 0 \end{bmatrix} \tag{15}$$

5.2. Smith predictor

Field tests performed at the diesel power plant revealed that there is a considerable dead time. For the tests described in this work, dead time is the time delay taken for the electric power signal starts to react after the application of a variation on the electro-hydraulic actuator. Smith Predictor is an adequate method to design controllers taking into account the observed delay [13]. The resulting controller with Smith Predictor consists of the digital controller $C(q^{-1}) = R(q^{-1})/S(q^{-1})$ with two additional internal feedback loops (see Figure 8), where one is a linearized estimated model of the plant without considering the time delay $\hat{G}_p(q^{-1})$, while the other takes into account the model with the time delay $q^{-d}\hat{G}_p(q^{-1})$.

6. Experimental field tests

The damping controller was successfully installed and validated by experimental field tests carried out at Santana Power Plant. The controller equipment has been installed in a cabinet of the control system of an 18 MVA diesel generating unit (Figure 9). Tests for model identification and estimation, with subsequent design and test of the proposed damping controller operating in closed-loop have been performed, with results showing an increase of the damping of the dominant electromechanical mode.

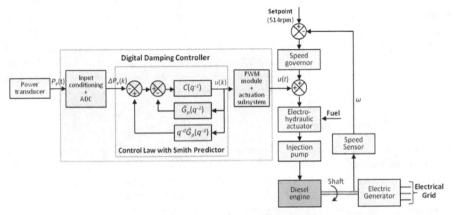

Figure 8. Digital controller with Smith Predictor.

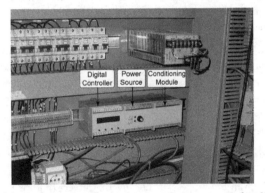

Figure 9. Field tests of the damping controller prototype, at Santana power plant [8].

6.1. Analysis of the system response for the application of a step variation

Step response is a useful way for an initial understanding of the system dynamic behavior, revealing some important characteristics that can be used for system modeling. Thus, with the damping controller disabled and the developed equipment programmed to operate only as a step generator, step variations were applied in the command of the fuel valve of the diesel engine. In consequence, the dominant oscillation mode with a frequency around 2.5 Hz, was observed in the electrical power signal. The amplitude of the test signal was configured to 5 mA, which is equivalent to an increase of 5% in the steady state opening of the fuel valve at the operation point considered, while the steady-state valve opening is set at 50 % (9 MW of active power).

These initial tests were also useful to identify possible nonlinearities of the system, such as deadband and time delay. These phenomena are usually found in hydraulic and thermal systems, such as the electro-hydraulic actuator and the diesel engine, and if they are ignored, may be difficult to tune the control system. In order to assess the nonlinearities, a series of tests was performed, which showed that the electro-hydraulic actuator has an excellent sensitivity and a dead zone that does not compromise the system control. So it was not necessary to implement any deadband compensation strategy in the controller. The field tests have shown that the diesel engine actuation system presents a dead time around 400 ms. In order to deal with this observed dead time, a Smith Predictor strategy was applied, as described in Section 5.2 of this chapter.

6.2. System identification tests

With the goal of obtaining a parametric model for the damping controller design, identification tests were performed in the 18-MVA diesel generating unit without using a damping controller. In order to excite the system electromechanical modes of interest, a PRBS sequence was used, with the parameter T_b equal to 80 ms and N equal to 9, resulting in a minimum frequency of 0.02 Hz and maximum frequency of 5.5 Hz, which excite uniformly the range of possible frequencies of the electromechanical oscillations modes (between 0.2 to 3 Hz) [1, 2]. The point of application of the exciting PRBS signal is the same point used for the application of the step response test, as already described in Section 6.1.

The input and output data sets were collected with a 40 ms sampling period and automatically transmitted to a PC, in which data sets were processed for purposes of identifying models representing the system dynamics in the current operating point. Figures 10a and 10b illustrate, respectively, the data obtained from the plant input variable (current variation in the diesel engine valve admission control) as well as from the plant output variable (power generator active power deviation).

From the acquired data, it is possible to make an estimate of the system response frequency spectrum on the operating point considered. The PRBS spectrum is characterized by being approximately uniform in the range specified by the project (Figure 10c), meaning that all modes between 0.1 and 5 Hz (approximately) were equally excited by the designed test signal. As can be seen in Figure 10d, a 2.5 Hz dominant intra-plant mode, having a small damping, can be observed from the electrical power deviation signal. Therefore, it is advisable to design a damping controller in order to improve the damping of this dominant oscillation mode.

The acquired data was divided into two data sets. The first one was used for the parametric model identification process, while the other set was used for model validating purposes. The identification process was carried out using a non-recursive least squares algorithm [10]. A fourth order ARX model structure was chosen, having 4 parameters in the numerator (B), 4 parameters in the denominator (A) and a discrete-time delay of 10 sampling intervals (400 ms).

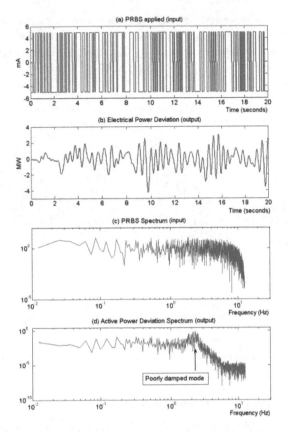

Figure 10. (a) PRBS applied in generator (b) Active power deviation of the generator (c) PRBS spectrum (d) Active power deviation spectrum [8].

In this way, the resulting identified plant input-output ARX model, which was used for control design, has the following structure:

$$q^{-d}\hat{G}_p\left(q^{-1}\right)=q^{-9}\frac{B(q^{-1})}{A(q^{-1})}=q^{-9}\frac{b_1q^{-1}+b_2q^{-2}+b_3q^{-3}+b_4q^{-4}}{1+a_1q^{-1}+a_2q^{-2}+a_3q^{-3}+a_4q^{-4}} \qquad (16)$$

where the parameters values are presented in the Table 3.

Parameter	a1	a2	a3	a4
Value	-1.9980	1.8254	-0.8676	0.2626
Parameter	b1	b2	b3	b4
Value	0.0033	0.0034	0.0030	0.0028

Table 3. Parameters of the model estimated for the plant ($T_s = 0.04$ s, $d = 10$ sampling intervals).

Figure 11 illustrates the comparison between the real output of the system and the output estimated using the fourth order model identified in the tests. The result shows the good estimation of the model parameters. It is also verified that the model successfully captured the oscillatory dynamic of the intra-plant electromechanical mode, observed in field tests performed in the diesel generating unit.

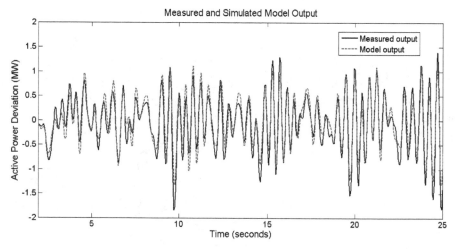

Figure 11. Comparison between the measured output (solid black) and the output of the model (dashed blue), for the diesel generating unit of 18 MW [8].

6.3. Design of the damping controller control law

Using the identified model, the damping part of the control law was obtained using the pole shifting method and solving the matrix equation system (15). The controller design main goal is to provide an acceptable damping (in this case, $\zeta_d = 0.2$), without affecting substantially the electromechanical dominant mode natural frequency and not exciting too much any unmodelled dynamics. In the plant fourth order ARX model (B / A), it is considered only one delay of sampling interval. The additional delay of 9 sampling intervals ($d=1+9$) is accommodated through the control law scheme implemented using the Smith Predictor, as explained in Section 5. Table 4 shows the parameters of the damping controller used in field tests.

Parameter	r0	r1	r 2	r3
Value	-1.6484	0.5728	-3.7522	-0.3732
Parameter	s1	s2	s3	-
Value	0.1018	0.0287	0.0020	-

Table 4. Parameters of the designed digital damping controller (T_s = 0.04 s).

6.4. Closed-loop control tests

After the controller design, its parameters were inserted into the embedded controller. Then the device was programmed to act in the closed loop control mode (damping controller mode), to provide an additional damping signal to the system, through the speed control loop of the diesel generating unit. To evaluate the performance of the damping controller, step type disturbances were applied, as can be observed during the tests illustrated in Figure 12.

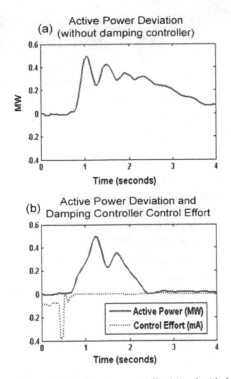

Figure 12. System step responses without damping controller (a) and with damping controller (b) [8].

Analyzing the graphics on Figure 13, it can be observed that the inclusion of the damping controller in the speed loop of Wärtsilä considerably improves the stability of the system, making it less oscillatory without affecting the speed governor performance.

In Figure 13 a comparison between the power spectral density of the active power deviation signal collected with the system with and without damping controller is shown. When the data were collected, the generator was excited by a PRBS signal. It is clear from this measurement that the observed electromechanical mode, around 2.5 Hz, is much more pronounced for the case in which the system operates without damping controller, than when the damping controller is acting on the system.

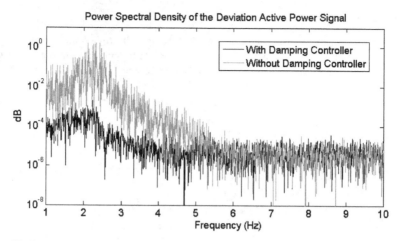

Figure 13. Comparison between the power spectral density of the deviation active power signals collected with damping controller and without damping controller [8].

7. Conclusion

This chapter presented the design and experimental tests of a digital damping controller which actuates through the speed governor system in order to damp an intra-plant electromechanical oscillation mode. The controller field tests were performed in a real generating unit of 18 MW in the Santana Thermoelectric Power Station.

The actuation of the damping controller through the speed governor was only possible due to the fast response of the diesel engine. Thus, this technology is not recommended for use in systems with machines that have low speed of actuation, as in generating units driven by hydraulic turbines.

Results demonstrated that the system performance was improved after the inclusion of the damping controller, thus increasing the system dynamic stability margins. No adverse interactions have been observed in all performed field tests. This task and others, such as, studies of an adaptive controller design, considering a variable time delay in the loop, will be object of future investigations.

Author details

Fabrício Gonzalez Nogueira*, Anderson Roberto Barbosa de Moraes,
Maria da Conceição Pereira Fonseca, Walter Barra Junior, Carlos Tavares da Costa Junior
and José Augusto Lima Barreiros
Federal University of Pará, Technology Institute, Faculty of Electrical Engineering, Belém, Pará, Brazil

* Corresponding Author

José Adolfo da Silva Sena, Benedito das Graças Duarte Rodrigues and Pedro Wenilton
Barbosa Duarte
Northern Brazilian Electricity Generation and Transmission Company (ELETRONORTE-Eletrobrás), Brazil

List of abbreviations

ARX	autoregressive with exogenous input
PWM	pulse width modulation
PSS	power system stabilizer
AVR	automatic voltage regulator
G_i	i-th generating unit
HMI	human machine interface
ADC	analog digital converter
UART	universal asynchronous receiver and transmitter
I/O	input/output
DC	direct current
DSC	digital signal controller
DSP	digital signal processor
PC	personal computer
PRBS	pseudo random binary sequence
LMS	least mean squares

List of symbols

T_s	data acquisition and control sample time
q^{-1}	discrete-time delay operator $\left(q^{-1}y(t) = y(t-1)\right)$
k	normalized discrete time
$B(q^{-1}), A(q^{-1})$	numerator and denominator polynomials of the plant ARX model
θ	vector of parameters
ϕ	data regressor vector
ε	estimating error
γ	step size of LMS algorithm
$R(q^{-1}), S(q^{-1})$	numerator and denominator polynomials of the digital controller
N, T_b	number of cells and sample generation interval for the shift register
f_{max}, f_{min}	maximum and minimal values for the desired excited frequency range
ω_n	dominant oscillation mode natural frequency
$A_{cl}(q^{-1})$	specified closed-loop polynomial
ξ_d	specified value of the damping for the dominant electromechanical mode
ξ	estimated value of the damping for the dominant electromechanical mode
α	pole-shifting factor
$\hat{G}_p(q^{-1})$	estimated model of the plant without considering the time delay
$\hat{d}$	estimated process discrete-time delay

$q^{-d}\hat{G}_p(q^{-1})$ estimated model of the plant considering the time delay
P_e active power signal
u control signal
ω rotor speed

Acknowledgement

The authors acknowledge the support from ELETRONORTE (Northern Brazilian Electricity Generation and Transmission Company), through the R&D project number 4500049067 (2005), and from CNPq (The Brazilian National Council of Research).

8. References

[1] Prabha S Kundur (1994) Power System Stability and Control. McGraw-Hill.
[2] Graham Rogers (2000) Power System Oscillations. Kluwer Academic Publishers Group.
[3] Wang H F, Swift F J, Hao Y S, Hogg B W (1993) Stabilization of Power Systems by Governor-Turbine Control, Electrical Power & Energy Systems, vol. 15, no. 6.
[4] Wang H F, Swift F J, Hao Y S and Hogg B W (1996) Adaptive Stabilization of Power Systems by Governor-Turbine Control. Electrical Power & Energy Systems, vol. 18, no. 2
[5] Yee S K, Milanović J V, Hughes F M (2010) Damping of system oscillatory modes by a phase compensated gas turbine governor. Electric Power Systems Research. vol. 80: 667-674.
[6] Wärtsilä (2007) Project guide Wärtsilä 46. Finland.
[7] Eletronorte. (2000) Field Tests to Perform Adjusts in the Parameters of the Speed Governor Controllers of the Santana Power Plant – Wärtsilä Generating Units. Technical Report (in Portuguese).
[8] Nogueira F G, Barreiros J A L, Barra Jr. W, Costa Jr. C T, Ferreira A M D. (2011) Development and Field Tests of a Damping Controller to Mitigate Electromechanical Oscillations on Large Diesel Generating Units. Electric Power Systems Research. 81 (2): 725-732.
[9] Ioan D. Landau, Gianluca Zito (2006) Digital Control Systems: Design, Identification and Implementation. 1. ed. Springer.
[10] Lennart Ljung, (1987) System identification: Theory for the user" University of Linköping Sweden, Prentice Hall, Englewood Cliffs, New Jersey, 1987.
[11] Paul Horowitz, Winfield Hill (1989) The Art of Electronics. Cambridge University Press, New York, USA, 2nd Edition.
[12] José A L Barreiros (1989) A Pole–Shifting Self Tuning Power System Stabilizer. MSc Thesis, UMIST, Manchester – UK.
[13] Karl J. Åström, Bjorn Wittenmark (1997) Computer Controlled Systems – Theory and Design, 3rd Edition, Prentice-Hall.

Chapter 10

Hardware-in-Loop Simulation Technology of High-Pressure Common-Rail Electronic Control System for Low-Speed Marine Diesel Engine

Jianguo Yang and Qinpeng Wang

Additional information is available at the end of the chapter

1. Introduction

1.1. Energy crisis and emission regulations

Oil price is increasing rapidly due to the oil reserve limited as a non-renewable resource. The proportion of the fuel cost rises in the total operating costs of ocean transportation companies. Compared with those from vehicles, the emissions of NOx and SOx from marine diesel engines are more serious, but the ones of CO_2, CO and HC are lower relatively. According to the statistics from International Maritime Organization (IMO), tens of millions tons of NOx and SOx are expanded into the atmosphere annually. An international protection regulation for the prevention and the control of marine pollution from ships— *Amendments to MARPOL Annex VI*—is adopted at Maritime Environment Protection Committee (MEPC) 58th meeting in October 2008. The detailed NOx emission targets of marine diesel engines are shown in Table 1. At present the techniques of energy conservation and emission reduction are the future technical developing direction of marine diesel engines and the research highlights for producers and institutes.

Rated revolution n (r/min)	n<130	1≤n≤2000	n>2000
Tier I (2000) g/kW*h	17.0	45*n-0.20	9.84
Tier II (2011)g/kW*h	14.36	44*n-0.23	7.66
Tier III (2016)g/kW*h	14.36	44*n-0.23	7.66
Tier III (2016)g/kW*h(ECA)	3.40	9*n-0.20	1.97

Table 1. IMO NOx emission standard

The primary technical methods to improve performance and to reduce emissions for marine diesel engine are shown below:

1. The density of intake air is increased through the turbocharging technology.
2. To raise the volumetric efficiency of the cylinder, the advanced techniques such as variable exhaust valve timing are used for improving the performance of intake and exhaust.
3. High pressure fuel injection is used to optimize fuel atomization and mix-ability of fuel and fresh air.
4. Fuel injection, mix and combustion are improved under variable working conditions by using suitable injection timing (VIT) based on fuel injection control system.

High-pressure common-rail (HPCR) electronic control technology is one of the most effective techniques to increase heat efficiency and to reduce emissions and noise for diesel engine. The fluctuation of the fuel injection pressure is not associated directly with diesel engine's speeds and loads. It means that the fuel injection pressure is independent in different conditions, especially the low-load condition. The performance parameters of the diesel engines, including excess air coefficient, combustion starting pressure, fuel atomization, are optimized through the flexible control of the fuel injection quantity, the injection plus width, the injection timing, the variable exhaust valve timing and etc. The fuel consumption and NO_x, PM emissions can be decreased substantially with the techniques implementation.

1.2. Hardware-in-loop simulation technology

Hardware-in-loop (HIL) simulation system is a close-loop dynamic testing system, composed of mathematic models and real-world physical parts. The mathematic models are established to simulate some other real-world physical parts which are non-existence or non-available for experiments in the system. HIL simulation technique is contributed to testifying the validity and feasibility of a designed proposal about the control system due to its flexible configuration. While some system settings change, the performance variation of the object measured can be observed simultaneously. Recently HIL simulation technique is widely used in developing the electronic control system of diesel engine, and it is acting as an important role during developing process. The benefits of HIL simulation system are shown in the following aspects:

1. A comprehensive emulation experimental environment is provided for the hardware and the control strategy certification. In addition, the experimental data can be as the evaluation index for the hardware being selected and matched.
2. The different working conditions of the engine can be simulated for testing software functions and control strategies. And a HIL simulation system can be applied to different projects due to its repeatability. The experimental cost on a HIL simulation system is reduced compared with that on a real diesel engine bench.
3. The initial calibration of electronic control system can be realized. The simulated engine model can be used to match and to demarcate basic control parameters to reduce developing period.
4. HIL simulation system is available for mass tests of system reliability.

5. When a electronic control system is tested by a HIL simulation system, it is easy to control and to record the testing process, and it is convenient to analyze and optimize the objects via calculating results from the models.

Normally, there are mainly four types (Figure 1) of HIL simulation system for different uses.

1. The first type HIL is mainly used in ECU developing and testing. The engine and mechanism are simulation models, but the ECU is real. This type
2. The second type HIL is mainly used in mechanism testing. The engine and ECU are simulation models, but the mechanism is real.
3. The third type HIL is mainly used in testing the combination of ECU and mechanism. The engine is simulation model, but the ECU and mechanism are real.
4. The fourth type HIL is mainly used in ECU developing and testing. The engine and mechanism are simulation models, but the ECU is real.

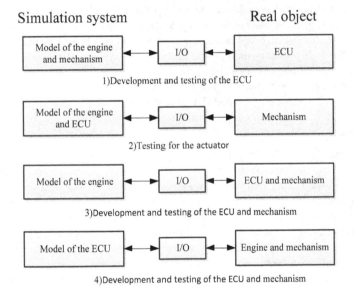

Figure 1. Application type of the simulation technique

The first type HIL cannot involve the entity of the engine and the mechanism, and the hardware costs of HIL are low relatively, besides the modeling methods are mature relatively, so the real time simulation technique is developing rapidly and widely used. The main corporations of real-time simulation system platform are dSPACE (Germany) , ADI (U.S.A) and NI (National Instrument, U.S.A) platform. Since 2005, NI corporation (National Instrument, U.S.A) has introduced PXI (PCI eXtensions for instrumentation) hardware platform with the RTOS (Real Time Operation System), which have combined its abundant I/O hardware resources, and have provided more choices to realize HIL real-time simulation system. There are many investigations on the first type HIL. F. R. Palomo Pinto realized HIL simulation of a fuel system using two PXIs with RTOS[1]. R. Isermann and J. Schaffnit realized HIL simulation system for ECU design and test using dSPACE[2]. Zhang Jie realized HIL simulation of high pressure common rail diesel engine on a common PC, using Linux RTAI[3]. The satisfactory effects have been obtained. The real-time feature of the model is required strictly in the first type HIL system.

As for the second type HIL, Bosch (Germany) in cooperation with EFS (Germany) developed a system testing platform, which has become more mature and widely used in recent years. According to the test requirements, users can set ECU model control parameters on the testing platform, such as the common rail pressure, the rail pressure control parameter, the pulse width of the fuel injection, the injection rate and so on. The research institutions and universities also successfully developed many testing platforms according to their research content and object. Wang Zhigang developed a middle pressure common rail diesel engine system testing platform[4]. Catania A. E. and Ferrari A. replaced the common rail ripe of Moehwald-Bosch testing platform by a high-pressure oil pipeline of smaller diameter. And the effect of the change to the fuel injection process is also accounted [5]. Shanghai Jiao Tong University and Shanghai Marine Diesel Engine Institute cooperated to develop the high pressure common rail fuel injection system reliability test platform[6]. The second type HIL has little demand of the engine models. ECU can only realize simple pressure control and electromagnetic valve control pulse output. Since there is no MAP of the fuel injection pressure and the fuel charge matching the engine load, it is impossible to combine the performance of the high pressure common rail system and the diesel engine load to research deeply on.

As for the third type HIL, Song Enzhe developed a semi-physical simulation platform according to type 16V396TE94 marine diesel engine and electronic control system[7]. The results of the experiments show that semi-physical simulation platform can be used for testing the diesel engine electronic control system in laboratory conditions, and the development cycle time can be shorten, and test costs can be reduced. Ou Dasheng designed a high-pressure common rail system test platform and carried out a series of tests on the platform[8], which includes the response characteristics of a high-speed solenoid valve, the hydraulic response of a electronically controlled fuel injector, the atomization characteristics of a high-pressure common rail system, the control of a common rail chamber pressure volatility, the common rail high pressure pump seal and seal ability, the solenoid valve fuel injector fuel injection characteristics and ECU performance tests. A higher reliability of the engine model is required in this kind of application.

The fourth type HIL is very close to a complete and original diesel engine. Shanghai Jiao Tong University designed a electronically controlled high pressure common rail diesel engine test bench system basing on GD-1 HPCR diesel engine, using the designed system to match and to calibrate the electronic control system of HPCR[9].

In summary, almost all the test benches of HPCR electronic control system are suitable for high and middle speed diesel engines, but there are rare reports and papers to introduce the test bench for the low-speed marine diesel engine.

2. HPCR electronic control system of low-speed marine diesel engine

2.1. Introduction

Nowadays the low-speed marine diesel engines with HPCR have successfully been put into commercial operation. These marine diesel engines with HPCR mainly include Wärtsilä RT-flex (Flex control) type marine diesel engine, MAN ME (Electronic control) type marine diesel engine and Japan Mitsubishi UEC Eco (emission control) type marine diesel engine. The first two corporations have accounted for more than 90% marine diesel engine with HPCR market shares.

2.2. RT-flex type marine diesel engine

The system architecture is shown in Figure 2. Comparing with a traditional diesel engine, RT-Flex marine diesel engine removed some components, such as exhaust valve drive devices, fuel pump, camshaft, servo motor for direction change, fuel connecting rod, starting air distributor and camshaft drive. These components are replaced with a fuel supply unit, a common-rail unit, an injection control unit, an exhaust control unit and WECS (Wärtsilä Engine Control System). WECS outputs 24V DC pulse signals to control the mechanical movement of the fuel injection and the exhaust valve behave via the high-speed solenoid valves and hydraulic forces. High-speed switching solenoid valves in the fuel injection control unit and the exhaust valve control unit are driven by the electromagnetic force, while hydraulic force is generated by high pressure servo oil in the servo oil common rail pipe. On the basis of the varied working conditions of the marine diesel engine, WECS can offer the flexible control of the injection start angle, the fuel injection pulse width and the open and the close angles of the exhaust valve, besides the mechanical delay is also fully considered. Multi-injection system (multi-jet system) is employed with the assembly of many injectors per cylinder. Three injectors were mounted with each cylinder in bore 600 mm and above of RT-flex marine diesel engine, but two injectors mounted with each

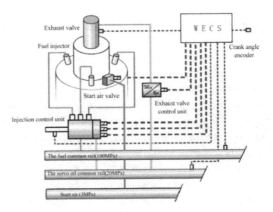

Figure 2. HPCR electronic control system of RT-flex

cylinder in bore 600 mm and below of RT-flex marine diesel engine. When the marine diesel engine starts, the start air valve on the upper cylinder is open and piston drives crankshaft rotating with the start air (3 MPa) via the 24V DC pulse control signals generated by WECS.

2.2.1. Oil supply unit

As is shown in Figure 3, the structure integrated is adopted in HPCR oil supply unit of RT-flex series marine diesel engine, with the features of a small footprint and easier installation and maintenance. The fuel pump, the servo oil pump, and the respective conduction-powered three-phase drive camshaft and the driving shaft are combined into a compact independent structure. The driving force for the oil supply unit is provided by gears set at the free end of the crankshaft. There is a fixed ratio between the speed of the camshaft, the crankshaft and the marine diesel engine crank (Speed of crank: Speed of Camshaft: Speed of the crankshaft=1:2.5:11.9).

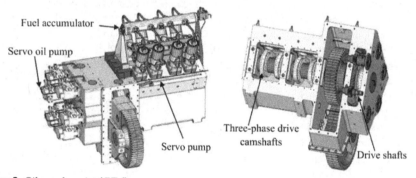

Figure 3. Oil supply unit of RT-flex

2.2.2. Common rail unit

As is shown in Figure 4, the common rail unit includes the fuel common rail pipe and the servo oil common rail pipe. The fuel common rail of 5RT-flex60C can afford the 100MPa pressure with the 5616 mm length and the peanut inwall. The operating pressure of the servo oil is the 20MPa with 5287 mm length and the round inwall. The fuel injection control unit is directly attached to the common rail fuel pipe, and the exhaust control unit is directly attached to the servo oil common rail. According to the fuel pressure signals of two pressure sensors set at the end of fuel common rail, the fuel pressure control signals are sent out to adjust the rotation angle of the fuel pump plunger cylinder from WECS, to accomplish a closed-loop control of the fuel common rail pressure. Similarly, a closed-loop control of the servo oil common rail pressure is still under WECS. Servo oil pressure signals are sent out by two pressure sensors set at the end of the servo oil common rail, and WECS adjust the servo pump swash plate tilt angle thus changing the fuel supply of the servo pump and accomplish a loop-locked control of the servo oil common rail pressure.

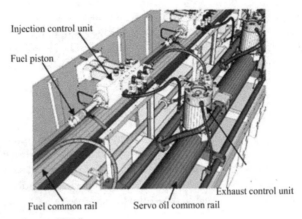

Figure 4. Common rail unit of RT-flex

2.2.3. Injection control unit

Figure 5 shows the injection control unit schematic structure of a single cylinder in no injection status. Injection rail valve converts to the open state under the excitation of the start pulse signals from WECS, then the high-pressure servo oil flow into the injection control valve. When the pressure produced by the servo oil is higher than the spring pre-tightening force, the state of the valve will change. The inlet oil line at the left side is closed, and the outlet oil line is opened. Because the pressure at the right side of the piston is higher than that at the left side, the piston is pushed to move to the left, meanwhile the high pressure fuel get into the injectors. The injection rail valve converts to the close state under the excitation of the close pulse signals from WECS, then the injection control valve will reset under the action of the spring. The outlet oil path at the left side of piston closes, while

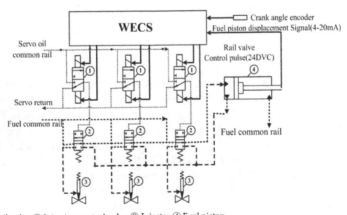

① Injection rail valve ② Injection control valve ③ Injector ④ Fuel piston

Figure 5. Principle of injection control unit

the inlet oil path opens. Although the pressure on both sides of the piston is approximately same, the discrepancy of the area is existed. The piston moves right back under the pressure differential, and the fuel pressure in the injector reduces. Based on the current crank angle and the actual fuel quantity calculated from the displacement of the fuel piston, WECS triggers the fuel injection order and correct the injecting angle.

2.2.4. Exhaust control unit

Figure 6 shows the principle of exhaust valve control unit when a single-cylinder exhaust valve is closed. It is mainly consist of ① exhaust rail valve, ② exhaust control valve, ③ isolated transmission device, ④ exhaust valve and ⑤ air spring and so on. According to the current crank angle, WECS triggers the opening and closing orders of the exhaust valves. The exhaust valve lift is used to detect the open\ close process and the open\close angle of the exhaust valve.

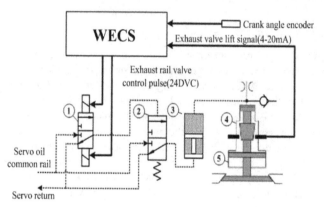

Figure 6. Principle of Exhaust Control Unit

2.2.5. ECU for RT-Flex diesel engine

WECS of RT-flex marine diesel engine includes six FCM-20 modules (Figure7). The FCM-20 module assembled on the cylinder is used to control start air valves, exhaust rail valves and fuel injection rail valves, and a single one is provided as online spare. Two FCM-20 modules of the cylinder 1 and the cylinder 2 are connected to AC20 speed control system. Three servo pumps are controlled separately by three FCM-20 modules of the cylinder 2, the cylinders 3 and the cylinders 4. And also a set of the fuel pumps (2-3per set) is controlled by two FCM-20 modules of the cylinder 3 and the cylinder 4. The functions of FCM-20 module are described in Table 2. Each FCM-20 has the same structure of hardware sub-systems. Some of them are indispensable for each cylinder, such as CYL-EU and VDM, while some of them are shared, such as COM-EU, MCM and CAN. Therefore, the functions and running parameters of each FCM-20 modules are set through the Flex-View software at the beginning configuration of WESC.

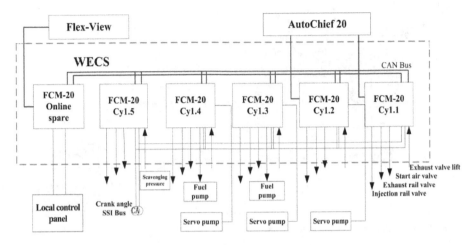

Figure 7. Structure of WECS

Number	Definition	Function
1	COM-EU	Communication and pressure control of the fuel common rail
2	ASM	Checking for the MCM
3	MCM	Communication and pressure control of the servo oil common rail
4	CAN	PWM signal output
5	CCM	Calculation for fuel quality, injection timing and exhaust valve timing
6	CYL-EU	Control for injection and exhaust valve
7	VDM	Amplifying the driven signal

Table 2. Function of the FCM-20 module

2.3. ME type marine diesel engine

ME marine diesel engine mainly consists of a hydraulic-machinery system for opening fuel injection valves and exhaust valves. The major executive components include two three-way solenoid valves, booster pumps, fuel injectors and electronic control exhaust valve, etc. As shown in Figure 8, there is only one high-pressure servo oil common rail in the HPCR system. The servo oil pump is driven by the diesel engine through the gear delivers high-pressure servo oil of 20MPa to the servo oil common rail. Then the high-pressure servo oil drives the booster pump and the exhaust valve piston through the solenoid valve. In the fuel oil system, each cylinder is assembled with a fuel oil pressure booster separately. With the help of the servo oil pressure, the fuel pressure rises from 1MPa to 75 - 120MPa. The solenoid valves control the movement of the plunger via the flow of the hydraulic oil, thus the controls of the fuel injection timing, the injection pressure and the fuel injection quantity are carried out.

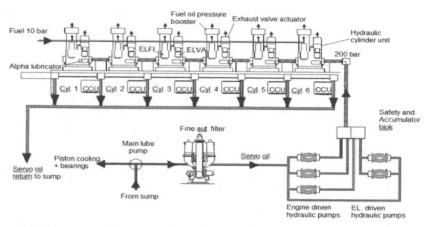

Figure 8. HPCR electronic control system of ME

2.3.1. Injection control unit

As is shown in Figure 9, the fuel injection control unit of ME-type marine diesel engine is called ELFI (Electronic Fuel Injection). The major component is a three way solenoid valve (NC Valve). The high-pressure servo oil flows into the fuel booster after the NC Valve opening. The area differential of the piston's two end caused high-pressure fuel injection. According to the fuel piston displacement, the actual fuel injection quantity is calculated and fuel injection angle is corrected in real time. The accurate control of the servo oil quantity can be achieved by the high-speed open and close action of NC valve. According to the different load conditions, it not only can achieve the change of the fuel injection timing angle and the cycle fuel injection quantity, but also can achieve fuel injection with the different fuel injection law.

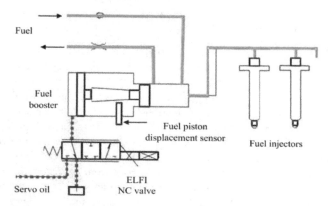

Figure 9. Injection control unit

2.3.2. ECU of ME diesel engine

As is shown in Figure 10, the electronic control system of ME diesel engine is divided into diverse function models in accordance with the different purposes.

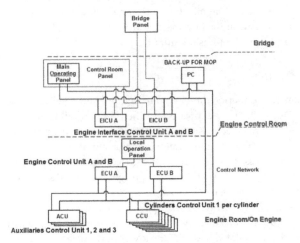

Figure 10. Structure of the electronic control unit

1. EICU (engine interface control unit) is responsible for interfacing with external system.
2. ECU (engine control unit) accomplishes the control functions of marine diesel engine, such as the speed, the operating mode and the starting sequence, etc.
3. ACU (auxiliary control unit) controls the hydraulic power supply unit of the pump and the auxiliary blower.
4. CCU (cylinder control unit) controls ELFI, ELVA (electronically controlled exhaust valve), SAV (air start valve), ALS (Alpha lubrication system) and HCU (hydraulic cylinder unit, including two parts of the fuel booster and exhaust valve actuator) of each cylinder.

Each module is installed in different parts of the marine diesel engine with the different control object and the function, as well as a set of backup module is set for the redundancy.

2.4. Comparative analysis between RT-Flex and ME

2.4.1. Similar points

In both types of the marine diesel engines, the mechanical control cams from traditional engine types are discard, which is used as the control core of the fuel injection and the exhaust valve timing, and replaced by the more precise and flexible ECU and electric-liquid-machine switching equipment[10]. The revolutionary improvement offers the low-speed marine diesel engine more flexibly features in adjusting control parameters, for instance the fuel injection timing, the fuel injection pulse width, the exhaust valve timing and etc. Air-fuel ratio, the pressure of the combustion starting point, the fuel atomization rate can be optimized, so the reduced fuel consumption and emissions are realized under the different conditions. Especially in the low-load working condition, the influence of NO$_x$ emission reductions is more obvious. The technique of high pressure fuel injection is adopted in the two types to improve the fuel atomization rate. The technique also guarantees the steady running of the diesel engine at the low speed and the low load. The minimum steady speed of the two types can reach 15 rounds per minute.

2.4.2. Dissimilar points

1. HPCR system of ME marine diesel engine is a pressure-charged system including the servo oil and the fuel. The fuel pressure of the common-rail is maintained at a medium level, for the pressure is set up by the booster oil pump. In the pressurized structure, the shape distortion of the common rail is small, and the sealing performance of the solenoid valve is less required, and the processing difficulty of exactitude parts is reduced. However, more fuel booster pumps are needed. The complexity and maintenance costs of the system increase.
2. HPCR system of RT-flex marine diesel engine also has two sets of oil ways. Fuel pressure is set up by oil supply units directly, so it is no necessary to boost pressure. The solenoid valves are used as the pilot valves with the lower flow rate and the response requirement. When the solenoid valves are opened, the piston is moved to make the fuel common rail and fuel atomizer connected. Because the pressure of the fuel common rail is kept at 100Mpa frequently, it is hard difficult to process the common rail.

3. Development of HIL simulation tested bench of HPCR electronic control system

3.1. Design of HIL simulation test bench of HPCR electronic control system

3.1.1. Structure of HIL simulation test bench

Various types of RT-flex series marine diesel engines are distinguished with the number of cylinders and the cylinder bore. There are not only 5 cylinders, 7 cylinders, 9 cylinders, 14 cylinders for RT-flex series marine diesel engines, but also the cylinder diameters cover 50 mm, 58 mm, 60 mm, 84 mm, 96 mm, etc. Although HPCR system structure, the length and the diameter of the common-rail pipe, the number of the fuel pumps and the servo oil pumps, and the number of injector per cylinder are slightly different for the specific type, the functions and principles are similar to each type including WECS and HPCR system, the common-rail technology with the fuel and the servo oil, the centralized supply unit for the high pressure fuel and the servo oil, the fuel injection control unit and the exhaust control unit, the system integration and the information communication technology. Those mentioned above are just the key parts that RT-flex type marine diesel engine is different with the traditional marine diesel engine. Based on the 5RT-flex60 CMKII HIL test bench is designed. The main technical parameters of the 5RT-flex60 CMKII are shown in Table 3.

Num.	Parameter	Unit	Value
1	Number of cylinder	—	5
2	Stroke	—	2
3	Bore	m	0.6
4	Pistonstroke	m	2.25
5	Rated power	kW	11800
6	Rated speed	r/min	114
7	Compression ratio	—	16.5
8	Fire order		1-4-3-2-5

Table 3. Main parameters of the 5RT-flex60CMKII

HPCR electronic control system of RT-flex series marine diesel engine is a cohesive whole according to the structure and function. The fuel supply unit, the common rail unit, the fuel injection control unit and the exhaust control unit are controlled by WECS. HIL simulation test bench is designed with integrated structure and the application type—Real-time simulation model of marine diesel engine, Control unit and executing mechanism. So the original structure characteristics and the functions of RT-flex series marine diesel engine are retained, and the system errors from the boundary conditions of the test bench are reduced. The structure integrated of HIL simulation test bench is divided into HIL simulation system and monitoring system [11], which is shown in Figure 11. Two parts are independent of each other, in addition the parameters and the data mainly are transferred via CAN bus. On HIL simulation test bench, the electronic control unit is WECS from the original machine, and the executive mechanism is properly simplified and improved based on the HPCR system of

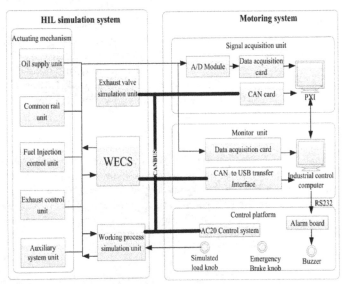

Figure 11. The structure of HIL simulation test bench

5RT-flex60C. Even more important, the real-time simulation model of the marine diesel engine which consists of the working process simulation unit and the exhaust valve simulation unit, is built up to keep the bench working well. The monitoring system of the bench is carried out for collecting, displaying and recording the test bench parameters. NI LabVIEW is served as the software development kit, besides the model and monitoring system are eployed separately in the compactRIO controller and PXI computer (from NI Corporation). The PXI is set in a console which including the Auto Chief 20 propulsion control system exporting the set point of the marine engine speed, and the load simulation device exporting physical signal for simulating the change of the engine load.

3.1.2. Functions of HIL simulation test bench

1.   Performance tests of HPCR electronic control system

The different working states of the diesel engine including starting, accelerating, decelerating, mutating and steady working can be simulated throughout HIL simulation test bench. So the bench can be used for the research on the control strategies and the working characteristics of HPCR electronic control system. All parts of the bench work together harmonically. The common-rail pressure, the fuel injection and the exhaust valve movement are controlled by WECS, while the different working conditions of the diesel engine are simulated by the real-time simulation model. The signals of the crankshaft angle, the fuel pump rack position, the fuel common-rail pressure, the servo oil common-rail pressure, the fuel injection quantity, the needle valve lift, the exhaust valve lift, the control from WECS are measured by the monitoring system.

2.   Fuel injection characteristic tests

The objective of the test is to conduct the dynamic working characteristics of the fuel control unit and injector. The fuel injection system is in running state independently, and some important parameter signals related to the fuel injection system, such as the crankshaft angle, the fuel common-rail pressure, the fuel injection quantity, the needle valve lift and reference injection signals from WECS are acquired and analyzed. However, the exhaust valves do not work, and the exhaust valve lift signals and various load working conditions of the diesel engine are simulated via the model.

3.   Exhaust valve characteristic tests

The objective of the test is to conduct dynamic working of the exhaust control unit and the exhaust valve. The exhaust system is in running state independently, and some important parameter signals which include the crankshaft angles, the servo oil common-rail pressure, the exhaust valve lift and the reference control are acquired and analyzed. However, the fuel injection system does not work, and the various load working conditions of the diesel engine are simulated via the model.

4.   Key executive mechanism tests

The key executive mechanisms of the high pressure common-rail system contain the fuel pump, the servo oil pump, the common-rail pipe, the fuel accumulator, the fuel injector, and the exhaust valve, etc. The performance tests of the parameters comparison for the executive mechanism can be proceed by applying the replacement way, also the executive mechanism can be tested independently.

3.2. Executive mechanisms of HIL simulation test bench

The sketch map of executive mechanisms is shown in figure 12. It is no feasibility that the same executive mechanisms from the 5RT-flex60C diesel engine are used for the test bench due to the huge structure. In the premise condition that the test bench can realize the integrated original system function, the fuel supply unit and the exhaust valve unit are simplified based on HPCR system. The tubing length, the diameter, the bending angle of the connector of the executive mechanism are close to the original machine as much as possible, and the key components which need to be focused on are retained. The common and non-key components are simplified and improved. An auxiliary system unit is added for providing the boundary conditions for the test bench. The auxiliary system unit is composed of the low pressure oil supply unit, compressed air unit, crank angle unit and fuel weighing unit.

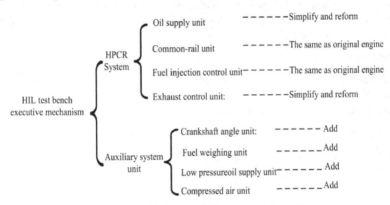

Figure 12. The sketch map of executive mechanism

1. Oil supply unit

The research on the fuel supply unit focuses on the oil supply way, the electronic control technology and the characteristics of the fuel pump and the servo oil pump, etc. The key components, which can reflect the functional structure characteristic, are kept such as the three-phase power cam and the fuel accumulator etc. To satisfy the function of HIL test bench and site layout, some simple components are replaced on the basis of the reliable equipment designed with the advanced and proven technique. The number of the fuel pump is 2, the one of which on the 5RT-flex60C diesel engine is 4. The number of the servo oil pumps is reduced from 3 to 1. Two sets of the variable-frequency adjustable-speed three-phase induction motors (be referred to as "fuel pump motor" and "servo oil pump motor") are served as the power of the oil supply unit. The speeds of motors are controlled by the real-time simulation model of the diesel engine, but the flow is controlled by WECS.

2.   Common-rail unit

The common rail unit including the fuel common rail and the servo oil common rail is conformity to that of the original machine.

3.   Fuel injection control unit

The fuel injection control unit is consistent with that of the original machine, which has one fuel injection control unit and three injectors per cylinder.

4.   Exhaust control unit

Due to the similarity of the exhaust valves movements, only one cylinder exhaust valve is retained. The exhaust control units installed on each cylinder are conformity to that of the original machine. Furthermore a 40L nitrogen bottle is used to provide the pressure for the air spring.

5.   Crankshaft angle unit

Crankshaft angle signal is provided for WECS by the crankshaft coder driven by an AC servo motor (be referred to as "crankshaft motor"). The motor speed can be controlled by the real-time simulation model of the diesel engine.

6.   Fuel weighing unit

Fuel weighing unit is two movable carriages comprised by three electronic balances and container. It is used to weighing the fuel injection quantity of three fuel injectors in the cylinder during a certain time or a certain cycle.

7.   Low pressure fuel supply unit and compressed air unit

Low pressure fuel supply unit is taken for recycling the backflow or the leakage fuel, the servo oil and the lubricating oil after multi-stage filtration. Compressed air unit provides compressed air of 0.5 ~ 0.75MPa by using Ingersoll-Rand Company UP5-15-7 type air compressor, to simulate the back gas pressure in the cylinder.

3.3. Real-time diesel engine simulation model

As the half of the working process simulation unit, the real-time diesel engine simulation model is used to simulate the diesel engine working process for coordinating synchronous operation of the test bench. The closed-loop system (Figure13) is composed by the real-time model, HPCR system, WECS and the console. The features of the real-time model are:

1.   To start three motors of the crankshaft angle, the fuel pump and the servo oil pump simultaneously through the enabling signals after detecting the control pulse signals of the starting air valves from WECS.
2.   To calculate the motors speed and scavenging pressure with the displacement signals of the fuel quality piston.
3.   To generate 4-20 mA signals to control the speed of three motors respectively.

4. To simulate the load and propulsion characteristics of the diesel engine with the load torque set on the console.
5. To generate two consistent simulated scavenging pressure signals to WECS.
6. To monitor the crankshaft angle and simulate backpressure in the exhaust valve through opening the pneumatic valve with the exhaust valve at the closing statue, and allow the compressed air go into a seal space at the bottom of exhaust valve.

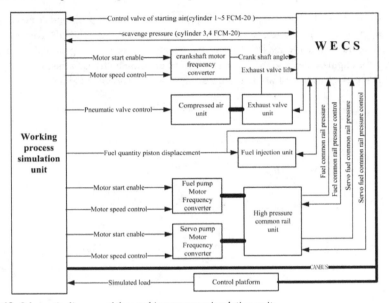

Figure 13. Schematic diagram of the working process simulation unit

Number	Crankshaft motor state	Control plus for staring air	Injecting	State
1	Stop	0V	No	Stopping
2	Stop	0V	Yes	—
3	Work	0V	Yes	Running stage
4	Work	0V	No	Stopping procedure
5	Stop	24V	No	Staring
6	Stop	24V	Yes	—
7	Work	24V	Yes	—
8	Work	24V	No	Staring procedure

Table 4. States of HIL simulation test bench

The executive mechanism status of HIL simulation test bench can be divided into eight forms shown in Table 4, according to crankshaft motor running, control pulse signal amplitude of starting air valve and the existence of injection process during a cycle of crankshaft motor. The different algorithms are designed to deal with different possible situation. Because injection order from WECS cannot be sent out when crankshaft motor stops or the starting air valve is open, the number 2, 6 and 7 will not appear. The others are the contents focused on.

3.3.1. Staring and staring procedure

At the staring stage of RT-flex marine diesel engine, the crankshaft position is calculated and the control pulse is outputted to open the starting air valve by WECS. Then the compressed air is blown into the cylinder to make the crankshaft rotate by pushing the piston. After a period of time, if the engine speed and the fuel common-rail pressure both reach the expected values, starting is success. On the contrary it is fail. Staring process is shown in Figure 14.

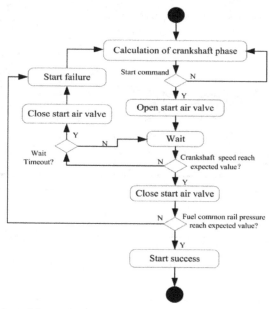

Figure 14. The flowchart of the starting

The primary control parameters in the process of diesel engine staring are shown in Table 5 and the opening angles of the starting air valve are shown in Figure 15, which come from experimental data of HIL simulation test bench.

Number	Category	Value
1	The waiting time(s)	15
2	The speed expected (r/min)	25
3	The pressure expected (MPa)	40

Table 5. The primary parameters in the starting process

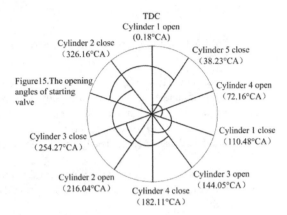

Figure 15. The opening angles of starting valve

As Figure16 is shown, the scavenging pressure (p_{sca}) is set 0.01MPa to simulate the real scavenging pressure of entity auxiliary fan. When the starting air valve control pulses are detected from any FCM-20 module, three motors will be started synchronously in the bound mode. However, in the non-bound mode only the crankshaft motor motor will be started.

With the curve fitting method, the crankshaft motor speed n_c (r/min) is regarded as the time logarithmic function according to the formula (1), and has to reach 25 (r/min) during 15 seconds. The current speed n_c' (r/min) is calculated by formula (2) according to the crankshaft angle difference ΔCA in 10 seconds. In the bound mode, the fuel pump motor speed n_f (r/min) and the servo oil pump motor speed n_s (r/min) are calculated respectively according to formula (3) and formula (4). Parameter a is the constant. Parameter t is the time (ms). Parameter n_f is the fuel pump bound parameter referring to chapter 2. Parameter n_s is the servo oil pump bound parameter referring to chapter 2.

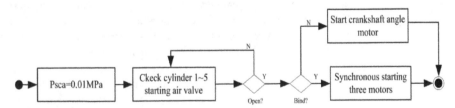

Figure 16. The control flowchart of the starting

$$n_c = \frac{\log_a(1+t)}{1000\times60} \tag{1}$$

$$n_c' = \frac{\Delta CA\times100\times60}{360} \tag{2}$$

$$n_f = k_f \times n_c' \tag{3}$$

$$n_s = k_s \times n_c' \tag{4}$$

3.3.2. Running stage

The simulation model of 5RT-flex60C at the running stage is restricted by the following conditions and design requirements.

1. The filling and emptying method can't be applied to the mathematical model of the diesel engine, because key parameters such as the intake/exhaust pipe volume, the scavenging volume and the turbocharger characteristics can not be obtained.
2. MAP of the measured data, such as MAP of the fuel pump supply rate and the diesel engine output power from the diesel engine bench, can not be used as the input boundary the model.
3. The purpose of the model at running stage is to stability control the speed of three motors, but not to predict the cylinder pressure, the temperature and the flow changing process of the intake and exhaust pipe.
4. The model needs to be interacted with WECS and the executive mechanisms in the real time. It is possible to be interfered by the external instable factors, such as the noise signals and speed fluctuations from motors, etc.
5. The changed working conditions of the diesel engine have to be simulated by adjusting the load simulation device in the HIL simulation test bench.
6. The understanding of WECS control strategies is limited, which includes the rail pressure, the fuel injection timing and the exhaust timing.

Therefore, the objective of the simulation model is to keep the test bench safe and reliable. The simulation model at the running stage is applied with the mean value engine model (MVEM), and the ventilation process model is simplified appropriately. MVEM is commonly used for describing the diesel engine behavior, and it is based on the global energy balance of the diesel engine. The algorithm of MVEM is very simple and reliable, and its resolving cost is very low. So that it is very suitable for the real-time application.

The propulsive characteristics of the diesel engine can be used to assess its performance, so the tests of the propulsive characteristics are adopted by the manufacturer, and an "official test report" with the propulsive characteristics of the real diesel engine based on the test bench is provided to the ship-owner before delivery, which contains the mass experimental data such as the speed, the scavenging pressure, the effective fuel consumption etc. under the working condition of 25% load, 50% load, 75% load, 90% load, 100% load, 110% load. Some experimental data from the official test report are used in the simulation model.

As shown in Figure17, MVEM is improved in basis of the restrictions mentioned above. It is supposed that the exhaust loss, the heat exchange loss and the mechanism loss of 5RT-flex60C diesel engine simulated in MVEM is the same with that of 7Rt-flex60C diesel engine, which has the same cylinder diameter and the stroke of RT-flex60C diesel engine. The specific fuel oil consumption (SFOC) at the standard ambient temperature from the "Official Test Report" of 7RT-flex60C diesel engine is calculated the diesel engine effective power, and the scavenging pressure from the "Official Test Report" of 7RT-flex60C diesel engine is as the load function with the curve fitting method.

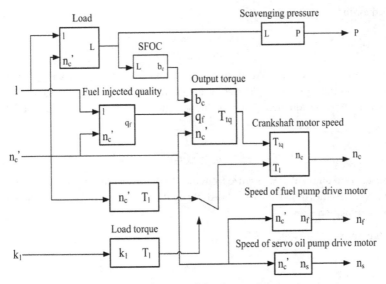

Figure 17. Schematic diagram of the simulation model at the running stage

The definition of the model boundary in this section refers to the following. The fuel quality displacement $l(mm)$ is acquired from the injection control unit. The current crankshaft motor speed n_c' (r/min) is calculated from formula (2). The load torque adjusting coefficient k_l is obtained from the load simulation device. The primary formulas in the improved model are as follows.

1. Load

$$L = \frac{n_c}{n_{MCR}} \times \frac{l}{l_{MCR}} \times 100\% \tag{5}$$

$l_{MCR}(mm)$ is the maximum displacement of fuel piston in the maximum continuous ratings (MCR) working condition. n_{MCR} (r/min) is crankshaft speed in MCR working condition.

2. Fuel injected quality per cycle

$$q_f = \begin{cases} \left(\frac{\pi \times r^2 \times l}{10^6} \times \rho - q_{leak}\right) \times N \\ \sum_{i=1}^{N}\left(\frac{\pi \times r^2 \times l_i}{10^6} \times \rho - q_{leak}\right) \end{cases} \tag{6}$$

$r(mm)$ is the radius of the fuel piston. $\rho(0.835kg/L)$ is the fuel density. N is the number of cylinder. $q_{leak}(kg)$ is the leak fuel quality per cycle, set as constant. The first equation in the equation (6) is used to calculate the fuel injected quality per cycle with the displacement of the fuel piston of any cylinder. And the second equation in the equation (6) is used to calculate the fuel injected quality per cycle with the displacement of fuel piston of N cylinders.

3.   Load torque

$$T_l = \begin{cases} k_l \times T_{MCR} k_l \in [0.2, 1.1] \\ (n'_c/n_{MCR})^2 \times T_{MCR} \end{cases} \tag{7}$$

T_{MCR} (N*m) is the load torque in MCR (maximum continuous rating) working condition. The first equation in the equation (7) is used in the condition of the load characteristic, and the second equation in the equation (7) is used in the condition of propulsion characteristics.

4.   SFOC

$$b_e = f(n'_c) \tag{8}$$

SFOC at standard ambient temperature from the "Official Test Report" of 7RT-flex60C diesel engine SFOC of the model is calculated throughout the formula (8) with the linear interpolation method.

5.   Output torque

$$T_{tq} = P_e \times \frac{9550}{n'_c} \tag{9}$$

$$P_e = \frac{60 \times n'_c \times q_f}{b_e} \tag{10}$$

P_e(kW) is the effective power.

6.   Scavenging pressure

$$p_{sca} = a_0 + a_1 L + a_2 L^2 + a_3 L^3 \tag{11}$$

a_0, a_1, a_2 and a_3 are constants.

7.   Crankshaft motor speed

$$n_c = n'_c + \frac{T_{tq} - T_l}{J} \times \frac{30}{\pi} \tag{12}$$

J(kg/m)is the rotational inertia of 5RR-flex60C diesel engine.

8.   Speeds of fuel pump motor and servo oil pump motor are calculated according to formula (3) and (4) respectively.

3.3.3. Stopping and stopping procedure

As Figure 18 is shown, the scavenging pressure is deceased to 0.01MPa in the stopping stage. The crankshaft motor speed is reduced by Δn_c(r/min) gradually. The speeds of the fuel pump motor and the servo oil pump motor are calculated according to formula (3) and (4) respectively. When the crankshaft motor speed is less than 5 (r/min), all of three motors are stopped in the bound mode, but in the non-bound mode only the crankshaft motor is stopped.

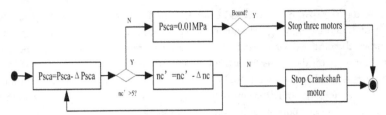

Figure 18. The control flowchart of the stopping

$$\Delta n_c = \Delta n_f / k_f \qquad (13)$$

Because the fuel pump motor has the biggest inertia moment, to keep three motors actual speed rate steady in stopping stage, Δn_c is calculated from the Δn_f *(r/min)* by the formula (13). Δn_f is the different speed of the fuel pump motor in the free decelerate condition.

3.4. Exhaust valve simulation model

The exhaust valve simulation model is as the other working process simulation unit. The main functions of the exhaust valve simulation model are to provide the real-time simulated exhaust valve lift signals for WECS. The exhaust valve lift signals are triggered by the exhaust valve opening and closing pulses from FCM-20 modules. Because of the difference between HIL simulation test bench and the original machine, the exhaust valve of the test bench cannot work properly, if the load of the test bench exceeds 75%. So two approaches are designed in the exhaust valve simulation model.

1. If the load of the test bench is below 75% load, the lift exhaust valve signals send to WECS, are from the real exhaust valve of the test bench. The lift exhaust valve signals of the real exhaust valve are sampled and saved by the exhaust valve simulation model, when exhaust valve opening/closing order is triggered. Then, the signals collected are

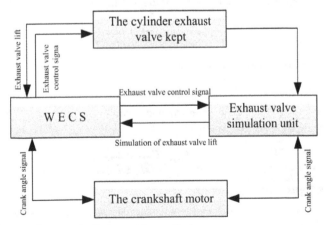

Figure 19. Schematic diagram of cylinder moving method

sent to the FCM-20 modules based on the "cylinder moving" method[13].The program flow of "cylinder moving" is shown in Figure 19. The "cylinder moving" method assumes that the working ways of all cylinders are all the same, and the working heterogeneity of the different cylinders is ignored. The working status of other cylinders is not directly calculated, but is obtained by the state recursive with the firing order. Therefore, not only the simulation speed is improved, but also the contradiction between the diesel engine's model accuracy and real-time is solved.

2. If the load of the test bench exceeds 75%, the emulator exhaust valve lift signals, simulated by the curve fitting method, are outputted by the exhaust valve simulation model. The emulator curve $l_{exh}(mm)$ is simplified for a trapezoidal, which is similar to the measured curve. The emulator curve is divided into four parts, including the closing status, the opening process, the opening status and the closing process[14]. The formulas for calculating the curve are shown below:

1. Closing status

$$l_{exh} = l_{min} \tag{14}$$

$l_{min}(mm)$is the displacement with minimum exhaust valve lift,.

2. Opening status

$$l_{exh} = l_{max} \tag{15}$$

$l_{max}(mm)$is the displacement with maximum exhaust valve lift.

3. Opening process

Simulated exhaust valve lift signals are calculated according to the formula (16), when exhaust valve opening order from FCM-20 module is delayed $t_{od}(ms)$.

$$\frac{dl_{exh}}{dt} = \begin{cases} \Delta l_{open} & (l_{exh} < l_{min}) \\ 0 & (l_{exh} \geq l_{min}) \end{cases} \tag{16}$$

$l_{open}(mm/ms)$is the exhaust valve opening rate.

4. Closing process

Simulated exhaust valve lift signals are calculated according to formula (17), when exhaust valve closing order from FCM-20 module is delayed $t_{cd}(ms)$.

$$\frac{dl_{exh}}{dt} = \begin{cases} -\Delta l_{close} & (l_{exh} > l_{min}) \\ 0 & (l_{exh} \leq l_{min}) \end{cases} \tag{17}$$

$l_{close}(mm/ms)$is the exhaust valve closing rate.

According to test analysis of exhaust valve, the initial values of model parameters are shown in Table 6.

Number	Parameter	Unit	Initial value	Number	Parameter	Unit	Initial value
1	l_{min}	mm	0	4	l_{max}	mm	73
2	t_{od}	ms	10	5	t_{cd}	ms	40
3	Δl_{open}	mm/ms	1.825	6	Δl_{close}	mm/ms	0.608

Table 6. Initial values of model parameters

3.5. Test verification

3.5.1. Starting process

HIL simulation test bench is started in the bound mode, and the start-up command comes from the Auto chief 20 system with the 51(*r/min*) setting speed. Figure 20 is shown that three motors start simultaneously and reach the setting speed at the same time in about 20 seconds. During the starting process, the three motors' speeds increase smoothly with the setting scale factors and achieve the desired objective.

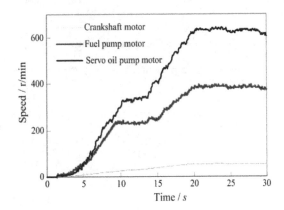

Figure 20. The motors speed in HIL simulation bench

3.5.2. Propulsion characteristic tests

The experimental data on HIL simulation test bench, including the crankshaft speed, the scavenging air pressure, the fuel consumption, the fuel indicator, the fuel rail pressure, the servo fuel rail pressure, the exhaust valve opening and closing angle, is compared with the values from the "Official Test Report" under the typical working conditions in propulsion characteristics test. The recorded curves of the crankshaft motor speeds in 5 seconds are shown in Figure 21(a). The speeds are relatively stable with less fluctuation. The speeds contrasts with the test bench and the report are shown in Figure 21(b). The relative errors of the speed from different data source are less than 1%.

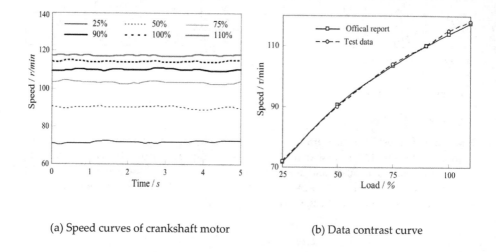

(a) Speed curves of crankshaft motor   (b) Data contrast curve

Figure 21. Speed curves of crankshaft motor in HIL simulation bench

Figure 22 is shown the scavenging air pressure. The relative errors between the test bench and report are less than 1.5%.

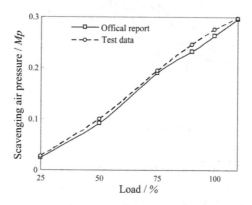

Figure 22. Data contrast curve of the scavenging air pressure

Figure 23 is shown the contrasts of the fuel consumption and fuel indicator. The relative errors of the fuel consumption are less than 1%, and the relative errors of the fuel indicator are less than 4%.

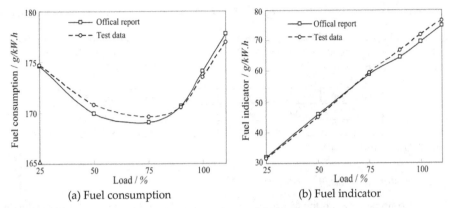

(a) Fuel consumption   (b) Fuel indicator

Figure 23. Data contrast curve of the fuel consumption and fuel indicator

Figure 24 is shown the contrasts of the exhaust valve opening and closing angles. The relative errors of opening angles are less than 0.2° CA (crank angle), and the relative errors of the closing angles are less than 0.1°CA.

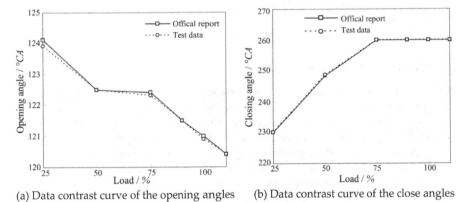

(a) Data contrast curve of the opening angles   (b) Data contrast curve of the close angles

Figure 24. Data contrast curve of the exhaust valve

Due to the existence of the engineering errors during diesel engine manufacturing, the performance data of the same type of low-speed marine diesel engine may be significantly different, let alone the measurement errors of the signals. Therefore, the existing errors on HIL simulation test bench are within the allowable range.

3.5.3. MCR full load shutdown test

During MCR full load shutdown test, the crankshaft motor speed is increased to the maximum instantaneous speed (n_{max}), then recoveries to MCR speed after an elapsed time (t_s) with the speed regulation (δ_1) according to the Formula (18).

$$\delta_1 = \frac{n_{max}-n}{n} \times 100\% \tag{18}$$

n_{max}, t_s and δ_1 in MCR full load shut down test on HIL simulation test bench is compared with the values provided in "Official Test Report" in Table 7, as a result the error is very small and could be allowed.

Number	Category	n_{max}/(r/min)	t_s/s	t_s
1	Official report	120	20.6	5.3
2	Test data	119.1	18.2	4.5

Table 7. Data contrast of the MCR

3.5.4. The minimum steady speed tests

The data contrast between the test bench and report is shown in Table 8. The data contain the diesel engine speed, the fuel indicator and the load in minimum steady speed test. As a result the errors are very small and could be allowed.

Number	Category	Speed/(r/min)	Fuel indicator/(%)	Load/(%)
1	Official report	16.0	12.5	3.5
2	Test data	15.3	13.0	2.7

Table 8. Data contrast of the minimum steady speed

4. Analysis of experiment result

The fuel common-rail pressure, the injection timing, the fuel injection pulse width and the fuel-injected quantity have an great influence on the fuel spray quality and the fuel injection law. Furthermore the combustion process and emissions of the diesel engines are also suffered the impact. The fuel injection mold of RT-flex diesel engine can be divided into three types, VIT ON (variable injection timing open), VIT OFF (variable injection timing off) and HEAVY SEA (diverse sea conditions). Based on the experimental data of the typical operating points in HIL simulation test bench, the control strategies of WECS are analyzed and investigated. The focus of analysis is the strategies of the fuel common rail pressure, the injection timing of VIT ON, VIT OFF and HEAVY SEA model, the control laws of the fuel injection pulse width and the fuel injected quantity, etc.

4.1. Injection control strategies in the low load condition

As shown in Figure 28, the tests are carried on in the HIL simulation test bench to figure out the control regulations of the injector in the low load condition. 25% load of the propulsive characteristics is set as the starting experiment point. Then, the load is decreased in accordance with the propulsive characteristics, and the injecting orders of the first cylinder from the FCM-20 module are measured. The load critical point, on which the number of the working injector is fallen from 3 to 2, can be determined with the amplitude variation. Additionally the test bench is kept operating in the current state for finding out the rotation law. In the same way the load critical point can be catch, on which the number of the working injector is fallen from 2 to 1. Similarly, the load critical points in the increased process of the load also can be found. The special laws and parameters are described as follows.

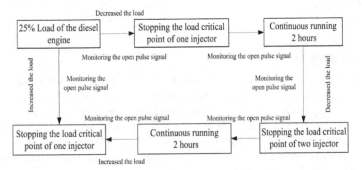

Testing scheme of the injection control strategy.

Figure 25. In the decreased process of the load, the test results show that when the load is down to 7%, the average alternating interval time is 1153.6s with the number of the working injector from 3 to 2. When the load is down to 3%, there is only one injector working. And the average alternating interval times change to 1153.4s. In the increased process of the load, 10% load is the turning point of the working injector number from 1 to 2. In the 15% load, the number of the working injectors recovers to 3. And the average alternating interval times change to 1154.5s.

4.1.1. Control strategies analysis of the fuel common rail pressure

4.1.1.1. Starting process

In the start-up phase, the common rail pressure is quickly established. In order to achieve the rapidity and stability, the open-loop control strategy is applied in WECS. The experimental curves of the fuel common rail pressure and control signals are shown in Figure 26. To ensure the maximum fuel delivery, the fuel pump control signal is kept with

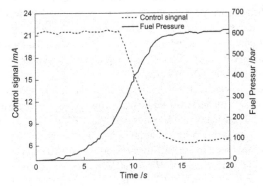

Figure 26. Curves of the fuel pressure and control signals

the 20 mA at the beginning, then the fuel pressure increases gradually with maintained control signal. When the actual pressures get to 25MPa, the control signal reduces to 7mA. While the actual pressures rise to the 60MPa, the control mode is translated into the PID (Proportion Integration Differentiation) closed-loop control state. To void pressure fluctuations, the current control signal is regarded as the initial value in the transition process.

4.1.1.2. Running process

The actual common rail pressures have to follow the target pressures at the engine working, so the closed-loop feedback control algorithm is used for the fuel rail pressure control. At the VIT ON and VIT OFF injection modes, the closed-loop control algorithm is used by WECS, and it is shown in Figure 27. The target pressures are get by looking up the rail pressure MAP chart according to the diesel engine load, and PID feedback control algorithm is carried out based on the difference value between the actual and the target pressures. What's more, the feed forward control is used to improve the system response performance.

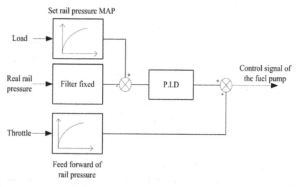

Figure 27. Closed-loop control algorithm

The actual pressure is unavoidable fluctuate in the fuel injection process. Also the pressure signals may be disturbed susceptibly. Therefore the signals need to be filtered to avoid the sharp pressure fluctuations caused by the mutations of control current signals. In addition, the target pressure may fluctuate wildly with the load changing when the diesel engine is working at the transient transition conditions. PID closed-loop control algorithm may result in a longer transition time of the actual rail pressure, which will impact on the fuel injecting and combusting. The feed forward control is added to the control algorithm to improve the response of the control system, and the feed forward control MAP of the common rail pressure is looked up via the fuel instruction. At VIT ON and VIT OFF injection modes, in the low load range of 0 to 15%, the target pressure value is 70MPa. In the load range of 15% to 25%, the target pressure value decreases to 60MPa. In the load range of 25% to 77%, the target value maintains 60MPa. In the load range of 77% to 90%, the target value gradually increases to 90MPa. In the load range more than 90%, the target value maintains 90MPa, and the feed forward control current signals is increased with the fuel indicate at the same percentage.

When the fuel injection mode is VIT ON or VIT OFF, the control strategies of the fuel common rail pressure prefer to reduce emissions at below 77% load, while prefer to improve the fuel economy when the load is more than 77% load. When the diesel engine load is less than 15% load, WECS will cut off parts of the injectors. Taking account of both the fuel economy and the emissions, the target value of the fuel common rail pressure is set as 70MPa. Since it can improve the combustion heat release rate, but not cause the NOx substantial increase. At the HEAVY SEA mode, PID closed-loop control algorithm is still active, but the target value of common rail pressure maintains 70MPa under various loads. The purpose is to avoid the actual pressure sharp fluctuation resulting in the mechanical components damaged.

4.1.2. Control strategies analysis of injection timing

The adjustment parameters of the injection timing angle will be freely set within a certain range by WECS according to the different fuel quality and the balance condition of each cylinder in the whole working situation. At VIT ON mode, in order to achieve optimal balance between economy and emissions of diesel engines, WECS adjusts the injection timing angle according to the scavenging pressure, the diesel engine speed and the change of the fuel common rail pressure. Variable injection timing angle is calculated as follows:

$$VIT \angle ° = inj.std + IT_{DEL} + FQS + VIT_A + VIT_B + VIT_c \tag{19}$$

The parameter $inj.std$ is set as the standard angle of the fuel injection timing, and the default value is 2 ° CA. The parameter IT_{DEL} is used for adjusting the imbalance working condition of each cylinder, caused by the tolerances of the manufacturing and the turbocharger matching. The parameter FQS is also the compensation of the fuel injection timing, which can be adjusted according to the fuel quality. If heavy oil is used, it will change the combustion lagging period of the diesel engine, then cause the cylinder pressure deviations in combustion process. The adjustment values of three injection timing mentioned above are set by user based on actual conditions. It will not change with the load in the operation process of the diesel engine. At VIT OFF and HEAVY SEA mode, the injection timing angle is the sum of three adjusted values.

At the VIT ON mode, VIT_A, VIT_B and VIT_c will be involved in the calculation of the variable injection timing angles. The parameter VIT_A is the fuel injection timing angle, which will be adjusted sectionally based on the scavenging pressure. When the scavenging pressure is lower than 0.35bar, the VIT_A value is set to 0. It is due to the low load of the diesel engine, and the auxiliary fan with the turbocharger does not work right now, so there is little significance to adjust the injection timing angle. When the scavenging pressure gradually increases from 0.35bar to 0.85bar, the value of VIT_A should gradually reduce from 0°CA to -2.5°CA. Since the ahead of the injection timing angle is benefited to improve the fuel economy at the low scavenging pressure. When the scavenging pressure continues rising to 0.85bar, the value of VIT_A should gradually rise from -2.5°CA to 1°CA. Since increased compression pressure and the delayed injection timing angle help to reduce NOx emissions with the low combustion temperature.

The parameter VIT_B is used to adjust injection timing angle according to the diesel engine speed. When the diesel engine load is the constant, the lower average effective pressure caused by the high speed results in the reduce of the combustion pressure. When the diesel engine speed is in the region of 70% to 100%, the value of VIT_B reduces gradually from $3°CA$ to $-1°CA$.

The parameter VIT_c is used to adjust the injection timing angle according to the fuel common rail pressure. The lower of the fuel common rail pressure would cause longer injection time and poor fuel atomization. The advancement of fuel injection timing angle through VIT_c is in favor of promoting combustion in order to improve the fuel economy of the diesel engine. When the fuel rail pressure increases gradually and exceeds the operating point pressure of MCR, the delay of fuel injection timing angle will compensate for the increased NOx emissions caused by too high fuel injection pressure via the VIT_c.

4.1.3. The main conclusions

1. If the load of diesel engine gradually reduces from high load to 7% or 3%, the number of the actual working injectors of each cylinder will reduce from three to two or one. If the load of diesel engine gradually rises from low load to 10% or 15%, the number of the actual working injectors of each cylinder will increase from one to two or three. The average alternating times of the injectors are approximate 1154s.

2. At VIT ON and VIT OFF mode, the control strategies of the fuel common rail pressure prefer to reducing emissions at the load below 77%. The control strategies prefer to improve the fuel economy when the load rises to more than 77%. WECS will cut off some of the injectors when the diesel engine load is less than 15% load. Taking account of both the fuel economy and the emissions, the target value of the fuel common rail pressure is set to 70MPa, and the objects are to improve the combustion heat release rate and avoid NOx increasing. The fuel injection mode of HEAVY SEA is set to prevent the mutations of the diesel engine load caused by adverse sea conditions. At the HEAVY SEA mode, the pressure of each load maintains 70MPa, and the PID closed-loop control algorithm is still used by WECS. Because the rough sea conditions could lead to the actual rail pressure sharp fluctuation resulting in the mechanical components damaged.

3. At VIT ON mode, when the diesel engine is working in low load working condition, fuel injection control strategies of WECS prefer to reducing emissions by taking a delay of injection timing angle. When the diesel engine is working in 75% load working condition, the engine speed is 90% of MCR speed, which is the commonly working condition of the actual operation of the marine diesel engine. Therefore, if the diesel engine is working at the 75% load or nearby, the fuel injection timing angle is set to advance to improve the fuel economy in priority.

4.2. Control strategies and characteristics of the exhaust valve system

4.2.1. Characteristics of the exhaust valve

The duration opening angles of the exhaust valve in difference working condition of the diesel engine are obtained in the test bench, which are calculated by the different angle between the corresponding angle at 15% full lift of the exhaust valve opening and the one at 85% full lift of the exhaust valve closing according to the experimental data. The duration opening angle increases along with the engine load. The open degree of the servo oil pump is taken to regulate the servo oil common rail pressure. And the open degree is controlled by the duty cycle of PWM (Pulse-Width Modulation) from WECS.

The delay times and angles of the exhaust valve opening are compared with the test data under different operating conditions. There is a hydraulic mechanism delay from the opening signal sending to the exhaust valve moving. When the exhaust valve is turned on, the spring is hit by the stem to produce the rebound from the spring. Along with the increase of the diesel engine load, the common rail pressure increases, but the delay time of the exhaust valve opening reduces. If the crank angle is set as the abscissa, the delay angle of the exhaust valve opening becomes large with the increase of the engine speed.

The delay times and the angles of the exhaust valve closing are compared with different operating conditions from the test data. There is a delay between the closing signal of the exhaust valve sent by ECU and the exhaust valve closing fully. The air spring pushes the exhaust valve until closed, and the delay time of the exhaust valve doesn't change significantly, but the speed and the time of delaying angle of the exhaust valve closing increase along with the working load. Therefore, the opening and closing of the exhaust valve are related to not only the system characteristics, but also the working conditions of the diesel engine.

4.2.2. The main conclusion

The setting angles of the exhaust valve opening and closing are confirmed by WECS based on the working condition of the diesel engine. The angles of the exhaust valve opening and closing are calculated through the measured lift curve of the exhaust valve. The control signals phase is adjusted by the difference between the setting value and the calculated one. The closed-loop control is used to make the exhaust valves opening / closing at a specified angle.

Some characteristics of control strategies of the exhaust valve are listed as follows[15]:

1. Servo oil common rail pressure rises together with the increasing of the working load of diesel engine.
2. The corresponding angle of the opening control signal of the exhaust valve reduces with the increasing of the working load of diesel engine.

3. The corresponding angle of the closing control signal of the exhaust valve tends to increase with the working load of the diesel increases from 25% load to 75% load, and then the angle descends.

4. The angle difference between the exhaust valve opening and closing increases with the working load of the diesel increasing from 25% load to 75% load, and then the angle descends.

5. The corresponding angle of the 15% full lift of the exhaust valve opening drops with the increasing of the working load of the diesel engine.

6. The corresponding angle of the 85% full lift of the exhaust valve closing increases with the working load from 25% to 75%, and the delay angle of the exhaust valve closing becomes large. So the corresponding angle is amended by WECS through reducing the control signal angle.

7. The different corresponding angles between the 15% full lift of the exhaust valve opening and the 85% full lift of the exhaust valve closing generally increase with the increasing of the working load of the diesel engine.

5. Conclusion

Hardware-in-loop simulation test bench of the HPCR electronic control system for low-speed marine diesel engine is developed basing on the Wärtsilä marine diesel engine. The working characteristics of the fuel injection system and the exhaust valve, and the control strategies under difference injection model are analyzed by the method of the experimental research together with simulation analysis. The main conclusions are listed as follows:

1. Typical structural and functional characteristics of two type low-speed intelligent marine diesel engine are analyzed. HPCR system of RT-flex marine diesel engine is the hydraulic-mechanical system which functions are independent but structure is inseparable from each other. The system is consisted of the oil supply unit, the common-rail unit, the fuel injection control unit, the exhaust control unit and so on. WECS is the control center of the HPCR system. The control signals and the rail pressure regulator signals from WECS impact on the working process and the state of the HPCR system directly. The common rail unit of ME marine diesel engine adopts the form of a single-cylinder assembled with a fuel supercharger. With the help of the servo oil pressure, NC valve can conduct the fuel injection with the various injection laws. The combustion process under difference working conditions is improved. The reduced fuel consumption and emissions will be achieved.

2. HIL simulation test bench is developed on the base of the analysis of the HPCR electronic control system of 5RT-flex60c marine diesel engine. The test bench contains the real time simulation model of the diesel engine unit and the monitoring system. The experiment results are shown that: The working characteristics of the HPCR system is in conformity with that of the original machine, and WECS control strategies are reflected distinctly in the test bench. The experimental conditions are provided for the research on the system characteristics, the control strategies and the performance of HPCR of high-power marine diesel engine.

3. Different typical working conditions of the propulsive characteristics of the diesel engine are simulated throughout the HIL test bench. The feature data of HPCR electronic control system are obtained, including the fuel pressure, the needle lift, the control pulse signals of the fuel injection, the exhaust valve opening and closing signals and the exhaust valve lifts and so on. Based on the experimental data, the investigations into control strategies of WECS, the characteristic feature of the fuel injection unit and exhaust valve control unit are focused on.

Author details

Jianguo Yang
School of Energy and Power Engineering, Wuhan University of Technology, P.R. China
Key Lab. of Marine Power Engineering &Technology (Ministry of Communications. P. R. China),
P.R. China

Qinpeng Wang
Key Lab. of Marine Power Engineering &Technology (Ministry of Communications. P. R. China),
P.R. China

6. References

[1] Pinto F. R. P., Vega-Leal A. P. (2010). A Test of HIL COTS Technology for Fuel Cell Systems Emulation. IEEE TRANSACTIONS ON INDUSTRIAL ELECTRONICS. C. 4(57): 1237-1244.

[2] R Isermann, J Schaffnit, S Sinsel (1999). Hardware-in-the-loop simulation for the design and testing of engine-control systems. Control Engineering Practice. J.7: 643-653.

[3] Jie Zhang (2007). A Dissertation Submitted in Partial Fulfillment of the Requirements for the Degree of Doctor of Philosophy in Engineering. Huzhuang University of Science and Technology. D

[4] Zhigang Wang (2006). Optimum Research on the Great Flow and Fast Response Electromagnetic Valve Used in the Medium Pressure Common Rail System of Diesel Engine. Wuhan University of Technology. D

[5] Catania A. E., Ferrari A. (2009). Further development and experimental analysis of a new common rail FIS without accumulator. American Society of Mechanical Engineers, Proceedings of the ASME Internal Combustion Engine Division Fall Technical Conference. C. p:219-228

[6] Ting Chen, Yuanming Gong, Bing Wang, Zhiyong Zhou (2009). Development of Reliability Test Platform for High-pressure Common Rail Fuel Injection System. Vehicle Engine. J.06:19-23

[7] Enzhe Song, Biling Song, Xiuzhen Ma (2010). Development of a semi-physical simulation platform for a marine diesel electronic control system. Journal of Harbin Engineering University. J.09:1153-1160

[8] DshengOu, Zhenming Liu, Jingqiu Zhang, GuangyaoOuyang (2008). Design and Experimental Study of a Comprehensive Test Platform for High-Pressure Common Rail System. Chinese Internal Combustion Engine Engineering. J.05:15-20

[9] YigangPeng, Hangbo Gong, Yuanming Gong, Xihuai Dai, Lin Yang, Bin Zhuo (2005). Design of Bench Testing System for Electronic Controlled High-Pressure Common-Rail Diesel Engines. Chinese Internal Combustion Engine Engineering. J.03:49-52

[10] Zhiang Wang, Jianguo Yang (2004). Analysis of Electronically-controlled Fuel System of MAN and Sulezer Diesel Engine. Diesel Engine. J. supplement:19-22

[11] Zheng Wang, Jianguo Yang, Wei Zhang (2010). Design and Implementation of HIL Simulation Test Platform of HPCR System for Intelligent Large Low Speed Diesel Engine. Chinese Internal Combustion Engine Engineering. J.06:54-58+64

[12] Zheng Wang, Jianguo Yang (2010). Method study for RT-flex HPCR low-speed diesel engine injecting law measuring. International Conference on Mechanical and Electrical Technology2010Singapore. C

[13] Qinpeng Wang, Xujing Tang, Zheng Wang, Chang Shu, Yonghua YU, Jianguo Yang (2009). ECU hardware in-loop simulation design for common rail system of diesel engine based on cRIO controller. Ship Engineering. J.05:13-16

[14] Qinpeng Wang (2009). The Design of Monitoring Control System with Heavy Duty Low-speed Diesel Engine. Wuhan University of Technology. D

[15] Jianguo Yang, Chang Shu, Qinpeng Wang (2012). Experimental research on exhaust valves based on a hardware-in-loop simulation system for a marine intelligence diesel engine. Journal of Harbin Engineering University. J.02:1-7

Permissions

The contributors of this book come from diverse backgrounds, making this book a truly international effort. This book will bring forth new frontiers with its revolutionizing research information and detailed analysis of the nascent developments around the world.

We would like to thank Dr. Saiful Bari, for lending his expertise to make the book truly unique. He has played a crucial role in the development of this book. Without his invaluable contribution this book wouldn't have been possible. He has made vital efforts to compile up to date information on the varied aspects of this subject to make this book a valuable addition to the collection of many professionals and students.

This book was conceptualized with the vision of imparting up-to-date information and advanced data in this field. To ensure the same, a matchless editorial board was set up. Every individual on the board went through rigorous rounds of assessment to prove their worth. After which they invested a large part of their time researching and compiling the most relevant data for our readers. Conferences and sessions were held from time to time between the editorial board and the contributing authors to present the data in the most comprehensible form. The editorial team has worked tirelessly to provide valuable and valid information to help people across the globe.

Every chapter published in this book has been scrutinized by our experts. Their significance has been extensively debated. The topics covered herein carry significant findings which will fuel the growth of the discipline. They may even be implemented as practical applications or may be referred to as a beginning point for another development. Chapters in this book were first published by InTech; hereby published with permission under the Creative Commons Attribution License or equivalent.

The editorial board has been involved in producing this book since its inception. They have spent rigorous hours researching and exploring the diverse topics which have resulted in the successful publishing of this book. They have passed on their knowledge of decades through this book. To expedite this challenging task, the publisher supported the team at every step. A small team of assistant editors was also appointed to further simplify the editing procedure and attain best results for the readers.

Our editorial team has been hand-picked from every corner of the world. Their multi-ethnicity adds dynamic inputs to the discussions which result in innovative outcomes. These outcomes are then further discussed with the researchers and contributors who give their valuable feedback and opinion regarding the same. The feedback is then collaborated with the researches and they are edited in a comprehensive manner to aid the understanding of the subject.

Apart from the editorial board, the designing team has also invested a significant amount of their time in understanding the subject and creating the most relevant covers. They scrutinized every image to scout for the most suitable representation of the subject and create an appropriate cover for the book.

The publishing team has been involved in this book since its early stages. They were actively engaged in every process, be it collecting the data, connecting with the contributors or procuring relevant information. The team has been an ardent support to the editorial, designing and production team. Their endless efforts to recruit the best for this project, has resulted in the accomplishment of this book. They are a veteran in the field of academics and their pool of knowledge is as vast as their experience in printing. Their expertise and guidance has proved useful at every step. Their uncompromising quality standards have made this book an exceptional effort. Their encouragement from time to time has been an inspiration for everyone.

The publisher and the editorial board hope that this book will prove to be a valuable piece of knowledge for researchers, students, practitioners and scholars across the globe.

List of Contributors

S. Jafarmadar
Mechanical Engineering Department, Technical Education Faculty, Urmia University, Urmia, West Azerbaijan, Iran

Minoru Chuubachi and Takeshi Nagasawa
Utsunomiya University, Japan

F. Portet-Koltalo and N. Machour
UMR CNRS 6014 COBRA, Université de Rouen, Evreux, France

Ulugbek Azimov
Department of Mechanical Engineering, Curtin University, Malaysia campus

Eiji Tomita and Nobuyuki Kawahara
Department of Mechanical Engineering, Okayama University, Japan

E.D. Banús, M.A. Ulla, E.E. Miró and V.G. Milt
Instituto de Investigaciones en Catálisis y Petroquímica, INCAPE (FIQ, UNL – CONICET), Santiago del Estero, Santa Fe, Argentina

Beñat Pereda-Ayo and Juan R. González-Velasco
Department of Chemical Engineering, Faculty of Science and Technology, University of the Basque Country UPV/EHU, Bilbao, Spain

Jungsoo Park
The Graduate School, Department of Mechanical Engineering, Yonsei University, Sinchon-dong, Seodaemun-gu, Seoul, Korea

Kyo Seung Lee
Department of Automotive Engineering, Gyonggi College of Science and Technology, Jeongwangdong, Siheung-si, Gyonggi-do, Korea

Daniel Watzenig
Graz University of Technology and Virtual Vehicle Research Center, Austria

Martin S. Sommer
Graz University of Technology, Austria

Gerald Steiner
Graz University of Technology, Austria

Fabrício Gonzalez Nogueira, Anderson Roberto Barbosa de Moraes, Maria da Conceição Pereira Fonseca, Walter Barra Junior, Carlos Tavares da Costa Junior and José Augusto Lima Barreiros
Federal University of Pará, Technology Institute, Faculty of Electrical Engineering, Belém, Pará, Brazil

José Adolfo da Silva Sena, Benedito das Graças Duarte Rodrigues and Pedro Wenilton Barbosa Duarte
Northern Brazilian Electricity Generation and Transmission Company (ELETRONORTE-Eletrobrás), Brazil

Jianguo Yang
School of Energy and Power Engineering, Wuhan University of Technology, P.R. China
Key Lab. of Marine Power Engineering &Technology (Ministry of Communications. P. R. China), P.R. China

Qinpeng Wang
Key Lab. of Marine Power Engineering &Technology (Ministry of Communications. P. R. China), P.R. China

Printed in the USA
CPSIA information can be obtained
at www.ICGtesting.com
JSHW011449221024
72173JS00004B/1004